Undergraduate Texts in Mathematics

Editors

S. Axler
F.W. Gehring
K.A. Ribet

Springer
New York
Berlin
Heidelberg
Barcelona
Budapest
Hong Kong
London
Milan
Paris
Santa Clara
Singapore
Tokyo

Undergraduate Texts in Mathematics

(continued after index)

Hugh Gordon

Discrete Probability

 Springer

QA 273 .G657 1997

Hugh Gordon
Department of Mathematics
SUNY at Albany
Albany, NY 12222
USA

Gordon, Hugh, 1930–

Discrete probability

Mathematics Subject Classification (1991): 60-01

Library of Congress Cataloging-in-Publication Data
Gordon, Hugh, 1930–
    Discrete probability / Hugh Gordon.
        p.   cm. — (Undergraduate texts in mathematics)
    Includes index.
    ISBN 0-387-98227-2 (hardcover : alk. paper)
    1. Probabilities.   I. Title.   II. Series.
  QA273.G657   1997
  519.2—dc21   97-10092

Printed on acid-free paper.

Production managed by Anthony Battle; manufacturing supervised by Jeffrey Taub.
Camera-ready copy prepared by the author.
Printed and bound by R.R. Donnelley & Sons, Harrisonburg, VA.
Printed in the United States of America.

9 8 7 6 5 4 3 2 1

ISBN 0-387-98227-2   Springer-Verlag New York Berlin Heidelberg   SPIN 10573550

*This book is dedicated
to my parents.*

Preface

Students of literature may have an advantage in understanding what this book is about. Noticing the many references to games and gambling, these students will immediately ask the right question: For what are the games and gambling a metaphor? The answer is the many events in the real world, some of them of great importance, in which chance plays a role. In almost all areas of human endeavor, a knowledge of the laws of probability is essential to success. It is not possible to work with actual uses of probability theory here. In the first place, applications in one area will be of limited interest to students of other areas. Second, in order to consider an application of probability in some field of study, one must know something about that field. For example, to understand how random walks are used by chemists, one must know the necessary chemistry. Since we do want to explain probability theory primarily by presenting examples, we choose to discuss simple things, like throwing a pair of dice. The examples stand in place of the many practical situations that are too complicated to discuss in this book.

Certainly the book is not intended as a guide to gambling, or even as a warning against gambling. However, before dropping the subject of gambling, a certain comment may be in order. We may as well quote the words used by DeMoivre, in his book, *The Doctrine of*

Chances, in similar circumstances. (Information about the persons quoted in this preface may be found in the body of the book.) Note that the doctrine of chances refers to what we now call probability theory. DeMoivre wrote, "There are many People in the World who are prepossessed with an Opinion, that the Doctrine of Chances has a Tendency to promote Play, but they soon will be undeceived, if they think fit to look into the general Design of this Book."

While applications abound, the present work can be read as a study in pure mathematics. The style is very formal, but we do have definitions, axioms and theorems; proofs of the theorems are given for those who care, or are obliged by an instructor, to read them. To quote DeMoivre again, "Another use to be made of this Doctrine of Chances is, that it may serve in Conjunction with the other parts of Mathematicks, as a fit Introduction to the Art of Reasoning; it being known by experience that nothing can contribute more to the attaining of that Art, than the consideration of long Train of Consequences, rightly deduced from undoubted Principles, of which this Book affords many Examples. To this may be added, that some of the Problems about Chance having a great Appearance of Simplicity, the Mind is easily drawn into a belief, that their Solution may be attained by meer Strength of natural good Sense; which generally proving otherwise, and the Mistakes occasioned thereby being not unfrequent, 'tis presumed that a Book of this Kind, which teaches to distinguish Truth from what seems so nearly to resemble it, will be look'd upon as a help to good Reasoning."

As we just noted, this book is about a part of mathematics—applicable mathematics to be sure, but the applications are not discussed here. Putting applications to one side, we look at the other side. There we find philosophy, which we also cannot discuss here: Is there such a thing as chance in the real world, or is everything predestined? Let us take a moment to at least indicate which topics we are not going to discuss.

One point of view was expressed by Montmort, writing in French in the eighteenth century. In translation what he says is, "Strictly speaking, nothing depends on chance; when we study nature, we are completely convinced that its Author works is in a general and uniform way, displaying infinite wisdom and foreknowledge. Thus to attach a philosophically valid meaning to the word "chance", we must believe that everything follows definite rules, which for the most part are not known to us; thus to say something depends on chance is to say its actual cause is hidden. According to this definition we may say that the life of a human being is a game where chance rules."

On the other hand, at the present time, there are those who claim that in quantum mechanics we find situations in which there is nothing but blind chance, with no causality behind it; all that actually exists before we make an observation is certain probabilities. We need not go into this point of view here, beyond noting the following: If we see from experiments that the rules of probability theory are followed in certain situations, for many purposes it is not necessary to know why they are followed. Be that as it may, the examples which appear in this book are all of the kind Montmort had in mind. We speak of whether a coin falls heads as a random event, because we do not have the detailed information necessary to compute from the laws of physics how the coin will fall.

* * *

How much of the book should be covered in a one semester course, or a two quarter course, is obviously up to the instructor. Likewise the instructor must decide how much to emphasize theorems and proofs.

Certain exercises are marked with two asterisks to indicate unusual difficulty. But who can say just how hard a problem is? What is certain is that the exercises as a whole were deliberately selected to cover a range of difficulties.

The book developed gradually over a long period of time. During that time the author was greatly aided by discussion of the work with his colleagues, many of whom suffered the misfortune of teaching using inadequate preliminary versions of the book; many flaws can to light this way. The author is most grateful for the help of these colleagues in the Department of Mathematics & Statistics of the State University of New York at Albany, especially George Martin, and for the help of the staff of that department.

Hugh Gordon

Contents

1

Introduction

Everyone knows that, when a coin is tossed, the probability of heads is one-half. The hard question is, "What does that statement mean?" A good part of the discussion of that point would fall under the heading of philosophy and would be out of place here. To cover the strictly mathematical aspects of the answer, we need merely state our axioms and then draw conclusions from them. We want to know, however, why we are using the word "probability" in our abstract reasoning. And, more important, we should know why it is reasonable to expect our conclusions to have applications in the real world. Thus we do not want to just do mathematics. When a choice is necessary, we shall prefer concrete examples over mathematical formality. Before we get to mathematics, we present some background discussion of the question we raised a moment ago.

Suppose a coin is tossed once. We say the probability of heads is one-half. The idea is, of course, that, in tossing a coin, heads occurs half the time. Before objecting to that statement, it is important to emphasize that it does seem to have some intuitive meaning. It is worth something; it will provide us with a starting point. To repeat, in tossing a coin, heads occurs half the time. Obviously, we don't mean that we necessarily get one head and one tail whenever we toss a coin twice. We mean we get half heads in a large number of tosses.

But how large is "large"? Well, one million is surely large. However, as common sense suggests, the chance of getting exactly five hundred thousand heads in a million tosses is small; in Chapter Six we shall learn just how small. (Meanwhile, which of the following do you guess is closest: 1/13, 1/130, 1/1300, 1/130000, 1/1300000000?) Of course, when we say "half heads," we mean approximately half. Furthermore, no matter how many times we toss the coin, there is a chance—perhaps very small—of getting heads every time. Thus all we can say is the following: In a "large" number of tosses, we would "probably" get "approximately" half heads. That statement is rather vague. Before doing anything mathematical with it, we must figure out exactly what we are trying to say. (We shall do that in Chapter Five.)

Now there are still additional difficulties with the concept of repetition. In 1693, in one of the earliest recorded discussions of probability, Samuel Pepys (1633–1703), who of course was a diarist, but not a mathematician, proposed a probability problem to Isaac Newton (1642–1727). When Newton began talking about many repetitions, Pepys reacted by rewording his question to make repetition impossible. (See Exercise 21 of Chapter Three for the reworded question.) Pepys was concerned with what would happen if something were done just once. (Curiously, Newton's letters to Pepys appear to be the earliest *documented* use of the idea of repetition in discussing probabilities. However, the idea must have been in the background for some time when Newton wrote about it. In fact, despite his great contributions in other areas, Newton does not seem to have done anything original in probability theory.) Another case where repetition is impossible occurs in weather forecasting. What does it mean to say the probability of rain here tomorrow is 60%? Even though there is considerable doubt in our minds as to exactly what we are saying, we do seem to be saying something.

Often the word "odds" is used in describing how likely something is to happen. For example, one might say, "The odds are 2 to 1 that it will rain tomorrow." Sometimes the odds given refer to the amount of money each party to a bet will stake. However, frequently it is not intended that any gambling take place. When that is so, making a statement associating an event with odds of 2:1 is simply a way of saying that of the two probabilities, namely, that that the event will happen and that that the event won't happen, one is twice the other. But which is twice which? Since "odds" basically refers to an advantage, and one side's advantage is the other's disadvantage, the matter is somewhat complicated. In addition, the use of negative words, for example, by saying "odds against," causes further difficulty. The meaning is usually clear from the exact wording and context, but no simple rule can be given. Under the circumstances, we shall never use the word "odds" again; the term "probability" is much clearer and more convenient.

Still before starting our formal theory, we begin to indicate the setting in which we shall work. We suppose that we are going to observe something that can turn out in more than one way. We may be going to take some action, or we may just watch. In either case, we speak, for convenience, of doing an experiment. We assume that which of the various possible outcomes actually takes place is "due to chance." In other words, we have no knowledge of the mechanism, if there is one, that produces one outcome rather than another. From our point of view, the way the experiment turns out is arbitrary, although some possibilities may be more likely than others, whatever that means. We restrict our attention to one experiment at a time. This one experiment often involves doing something more than once. While, over the course of time, we consider many different examples of experiments, all our theory assumes we are studying a single experiment. The basic idea of probability theory is that we assign numbers to hypothetical events, the numbers indicating how likely the events are to occur. Clearly we must discuss what an event is before we can consider assigning the numbers. From now on, when we call something an event, we do not mean that it has already taken or necessarily will take place. An event is, informally, something that possibly could occur and for which we want to know the chances that it will occur. A more formal, and more abstract, definition of an event occurs in the next paragraph.

From the point of view of abstract mathematical theory, a sample space is just a set, any set. Since we are going to study only a certain portion of probability theory, which we shall call *discrete probability*, we shall very soon put a restriction on which sets may be used. However before doing anything else, we want to describe the intuitive ideas behind our mathematical abstractions. We make a negative comment first: The word "sample," which we use for historical reasons, does not give any indication of what we have in mind. The *sample space*, which we shall always denote by Ω, is simply the set of all possible ways our experiment can turn out. In other words, doing the experiment determines just one element of Ω. Thus, when the experiment is done, one and only one element of Ω "happens." The choice of a sample space then amounts to specifying all possible outcomes, without omission or repetition. A subset of the sample space is called an *event*; we shall assign probabilities to events. We illustrate the concepts of sample space and event in the following examples:

1. A die is to be thrown. We are interested in which number the die "shows"; by the number shown, we mean, of course, the number on the top of the die after it comes to rest. A possible choice of sample space Ω would be $\{1, 2, 3, 4, 5, 6\}$. Other choices of Ω are also possible. For example, we could classify the way the die falls considering also which face most nearly points north. However, since we are interested only in which face of the die is up, the simplest choice for Ω would be $\{1, 2, 3, 4, 5, 6\}$, and we would be most unlikely to make any other choice. We consider some examples of events: Let A be "The die shows an even number." Then A is an event, and $A = \{2, 4, 6\}$. Let B be "The die shows a number no greater than 3." Then B is an event, and $B = \{1, 2, 3\}$. Let C be "The die shows a number no less than 10." Then C is an event, and $C = \emptyset$, the empty set. Let D be "The die shows a number no greater than 6." Then D is an event, and $D = \Omega$. (We note in passing that $\emptyset$ and Ω are events for any experiment and any sample space.)

2. Suppose two dice, one red and one blue, are thrown. Suppose also that we are concerned only with the total of the numbers shown on the dice. One choice of a sample space would be $\Omega = \{2, 3, 4, \ldots, 11, 12\}$. Another choice for Ω would be the set

$$\begin{aligned}
\{&(1, 1), (1, 2), (1, 3), (1, 4), (1, 5), (1, 6), \\
&(2, 1), (2, 2), (2, 3), (2, 4), (2, 5), (2, 6), \\
&(3, 1), (3, 2), (3, 3), (3, 4), (3, 5), (3, 6), \\
&(4, 1), (4, 2), (4, 3), (4, 4), (4, 5), (4, 6), \\
&(5, 1), (5, 2), (5, 3), (5, 4), (5, 5), (5, 6), \\
&(6, 1), (6, 2), (6, 3), (6, 4), (6, 5), (6, 6)\},
\end{aligned}$$

with 36 elements; here the notation (a, b) means the red die shows the number a and the blue die shows the number b. The first choice has the advantage of being smaller; we have only 11 points in Ω instead of 36. The second choice has the advantage that the points of Ω are equally likely. More precisely, it is reasonable to assume, on the basis of common sense and perhaps practical experience, that the 36 points are equally likely. At this time, we make no decision as to which Ω to use.

3. Three coins are tossed; we note which fall heads. The first question is whether we can tell the coins apart. We shall always assume that we can distinguish between any two objects. However, whether we do make the distinction depends on where we are going. In the case of the three coins, we decide to designate them as the first coin, the second coin, and the third coin in some arbitrary manner. Now we have an obvious notation for the way the coins fall: hht will mean the first and second coins fall heads and the third tails; tht will mean the first and third coins fall tails and the second heads; etc. A good choice of a sample space is

$$\Omega = \{hhh, hht, hth, htt, thh, tht, tth, ttt\}.$$

By way of example, let A be the event, "Just two coins fall heads." Then $A = \{hht, hth, thh\}$. The event, "The number of tails is even" is $\{hhh, htt, tht, tth\}$. It is worth noting that it would not have mattered if we tossed one coin three times instead of three coins once each.

4. A coin is tossed repeatedly until it falls the same way twice in a row. A possible sample space, with obvious notation, is

$$\Omega = \{hh, tt, htt, thh, hthh, thtt, hthtt, ththh, \ldots\}.$$

Examples of events are "The coin is tossed exactly five times" and "The coin is tossed at least five times."

In studying probability, it makes a great deal of sense to begin with the simplest case, especially since this case is tremendously important in applications. In certain situations that arise later, we shall use addition where more complicated examples require integration. The difference between addition and integration is precisely the difference between discreteness and continuity. We are going to stick to the discrete case. We want the points of our sample space to come along one at a time, separately from each other. We do not want to deal with the continuous flow of one point into another that one finds, for example, for the points on a line. In more precise terms, we restrict ourselves to the following two possibilities. One case we do consider is the one where the sample space contains only finitely many points. In fact, the majority of our examples involve such sample spaces. Sometimes we consider sample spaces in which the points can be put into one-to-one correspondence with the positive integers. In other words, we always assume that, if we so choose, we can designate the points of our sample space as the first point, the second point, and so on, either forever or finally reaching some last point. Note that we said the points can be arranged in order, not that they *are* arranged in order. In other words, no particular order is presupposed. From now on, sample space will mean any set with the property just announced.

The main reason for using sets in probability theory is that the set operations of union and intersection correspond to the operations "and" and "or" of logic. Before describing that correspondence, let us clarify our language by considering a single event A. When we do our experiment, the result is just one point u of the sample space; in other words, u occurs. To say that the event A happens is to say that that $u \in A$. Now we turn to the event $A \cup B$. $u \in A \cup B$ means that either $u \in A$ or $u \in B$. Thus $A \cup B$ happens exactly when either A or B happens. (Note that, in conformity with standard mathematical usage, "either A or B happens" does not exclude the possibility that both happen.) Likewise, the event $A \cap B$ is the event, "A and B both happen."

If A is an event, $\overline{A}$ denotes the event "A does not happen." In set theoretic terms, $\overline{A}$ is the set of those $u \in \Omega$ such that $u \notin A$. Of course, we may use the bar over any symbol that denotes an event; for example, $\overline{A \cup B}$ means the event, "$A \cup B$ does not happen."

Finally, the notation $A \setminus B$ denotes the event, "A happens, but B does not." In other words, $A \setminus B = A \cap \overline{B}$. In terms of sets, $u \in A \setminus B$ if and only if both $u \in A$ and $u \notin B$.

Let us explore, for practice, the way the notation just introduced works out in slightly more complicated cases. For example, consider $C = \overline{A \cap B}$. To say that C happens is to say that A and B do not both happen; in other words, that at least one of A and B fails to happen.

Thus, $C = \overline{A} \cup \overline{B}$. The logic here is, of course, the logic behind the set theoretic identity $\overline{A \cap B} = \overline{A} \cup \overline{B}$. Likewise one can interpret, in terms of events, the identity $\overline{A \cup B} = \overline{A} \cap \overline{B}$. [These last two equations are called DeMorgan's Laws, after Augustus DeMorgan (1806–1871).]

Often we have two ways to say the same thing, one in the language of sets and one in the language of events. Suppose A and B are events. We can put the equation $A \cap B = \emptyset$ in words by saying that the sets A and B are *disjoint*. Or we can say equivalently that A and B are mutually exclusive; two events are called *mutually exclusive* when it is impossible for them to happen together. Consider the set theoretic identity, which we shall need soon, $A \cup B = A \cup (B \setminus A)$. A logician might seek a formal proof of that identity, but such a proof need not concern us here. We convince ourselves informally that the equation is correct, whether we regard it as a statement about sets or about events, just by thinking about what it means. Additional equations that will be needed soon are $A \cap (B \setminus A) = \emptyset$, $(A \cap B) \cup (A \setminus B) = A$ and $(A \cap B) \cap (A \setminus B) = \emptyset$. In each case, the reader should think over what is being said and realize that it is obvious.

Now the key idea of probability theory is that real numbers are assigned to events to measure how likely the events are. We have already discussed this idea informally; now we are ready to consider it as a mathematical abstraction. We begin by preparing the groundwork as to notation. The probability of an event A will be denoted by $P(A)$; in conformity with general mathematical notation, $P(A)$ is the value the function P assigns to A. In other words, P is a function that assigns numbers to events; $P(A)$ is the number assigned by P to A.

Certain events contain just one point of the sample space. We shall call such an event an *elementary* event. Consider Example 3 above, where three coins are tossed. The event, "Three coins fall heads" is an elementary event, since it consists of just one of the eight points of the sample space. This event is {hhh}, and its probability will be denoted by $P(\{hhh\})$, according to the last paragraph. To simplify notation, we shall write $P(hhh)$ as an abbreviation for $P(\{hhh\})$. In general, if $u \in \Omega$, $P(u)$ means $P(\{u\})$.

We shall need to refer to the sum of the numbers $P(u)$ for all $u \in A$, where A is an event. We denote this sum by

$$\sum_{u \in A} P(u).$$

If A contains finitely many points, but at least two points, then there is no problem in knowing what we mean; we simply add up all the numbers $P(u)$ corresponding to the various points $u \in A$. If A contains a single point u, we understand the sum to be the one number $P(u)$. If $A = \emptyset$, we follow the standard convention of regarding the sum as zero. If A contains infinitely many points, there is an easily solved problem. The assumption made when we discussed discreteness above guarantees that we can arrange the numbers we want to add into an infinite series. Since all the terms are nonnegative, it makes no difference in which order we put the terms. The advantage of the notation

$$\sum_{u \in A} P(u)$$

is that it covers all these cases conveniently.

We have already said that P is a function that assigns a real number to each event. We now suppose

1. $P(A) \geq 0$ for all events A.

2. $P(\Omega) = 1$.

3. For every event A,

$$P(A) = \sum_{u \in A} P(u);$$

in particular, $P(\emptyset) = 0$.

Some consequences of these assumptions are too immediate to be made into formal theorems. We simply list these statements with brief parenthetical explanations; throughout, A and B denote events.

a. If $A \subset B$, then $P(A) \leq P(B)$. ($P(A)$ is the sum of the numbers $P(u)$ for all $u \in A$; $P(B)$ adds in additional nonnegative terms.)

b. $P(A) \leq 1$. (This is the special case $B = \Omega$ of the last statement.)

c. If A and B are mutually exclusive, then $P(A \cup B) = P(A) + P(B)$. ($P(A)$ is the sum of all $P(u)$ for certain points u of Ω; $P(B)$ is the sum of all $P(u)$ for certain other points of Ω. $P(A \cup B)$ is the overall sum of all of these $P(u)$.)

d. $P(\overline{A}) = 1 - P(A)$. (This amounts to the case $B = \overline{A}$ of the last statement, since $P(A \cup \overline{A}) = P(\Omega) = 1$.)

Statement c can obviously be generalized to more than two events. We call events $A_1, A_2, \ldots, A_n$ or events $A_1, A_2, \ldots$ mutually exclusive if A_i and A_j are mutually exclusive whenever $i \neq j$. If $A_1, \ldots, A_n$ are mutually exclusive, then

$$P(A_1 \cup \cdots \cup A_n) = P(A_1) + \cdots + P(A_n).$$

If $A_1, A_2, \ldots$ are mutually exclusive, then

$$P(A_1 \cup A_2 \cup \cdots) = P(A_1) + P(A_2) + \cdots;$$

note that the sum on the right is an infinite series. The verification of this last statement about a series requires a deeper understanding of the real number system than the other statements, but nevertheless we shall occasionally, although not frequently, make use of this conclusion.

By way of variety, let us make the next statement a formal theorem. This theorem gives us a formula for $P(A \cup B)$ that works even when A and B are not mutually exclusive.

Theorem *Let A and B be events. Then*

$$P(A \cup B) = P(A) + P(B) - P(A \cap B).$$

Proof Since A and $B \backslash A$ are mutually exclusive and $A \cup B = A \cup (B \backslash A)$,

$$P(A \cup B) = P(A) + P(B \backslash A).$$

Since $A \cap B$ and $B \backslash A$ are mutually exclusive and $B = (A \cap B) \cup (B \backslash A)$,

$$P(B) - P(A \cap B) + P(B \backslash A).$$

Combining the last two equations by subtraction we have

$$P(A \cup B) - P(B) = P(A) - P(A \cap B).$$

Adding $P(B)$ to each side, we have the desired conclusion.   □

Most, but not all, of the sample spaces we shall be using in this book are of a special kind. They are sample spaces that consist of finitely many "equally likely" points. A little more formally, assume that all elementary events have the same probability. In other words, there is a number a such that $P(u) = a$ for all $u \in \Omega$. Denote the number of points in Ω by N. Now we have

$$1 = P(\Omega) = \sum_{u \in \Omega} P(u) = a + \underset{N \text{ terms}}{\cdots} + a = Na.$$

Thus a must be $1/N$. Next consider an arbitrary event A, and suppose there are M points in A. Then we have

$$P(A) = \sum_{u \in A} P(u) = a + \underset{M \text{ terms}}{\cdots} + a = Ma.$$

Thus $P(A) = Ma = M(1/N) = M/N$. Thus, the probability of A can be determined by counting the number of points in each of A and Ω.

The formula just given has often been used as a definition of probability. When one does that, one says the probability of an event is the number of "favorable cases," that is, cases in which the event occurs, divided by the total number of possible cases. It must be stated explicitly or merely understood that the cases are equally likely. We prefer the abstract definition we used above partly because of its greater generality.

In the situation just discussed, finding probabilities is a matter of counting in how many ways certain things can happen. Of course, if the sample space is at all large, we cannot do the counting by writing out all the possibilities. In the next chapter, we shall learn some basic techniques for finding the number of elements in a set.

How do we know when the points of the sample space are equally likely? The fact is that we don't know unless someone tells us. From the point of view of mathematics, we must begin by clearly defining the problem we are working. To save time in stating problems, we agree once and for all to use symmetry as a guide; in other words, we assume events that are "essentially identical" to be equally likely. We use common sense in deciding what "essentially identical" means. A few examples will clarify matters. When we say a die is thrown, we mean the obvious thing; a die with six faces is thrown in such a manner that each face has an equal chance to come out on top. If the die were loaded, we would say so. If the die had, for example, eight faces, we would say so. When we speak of choosing a ball at random from an urn, we mean that each ball in the urn has the same chance to be chosen. On the other hand, suppose we put in an urn two red balls and three blue balls. Then a ball is drawn at random. We do not assume that each color has an equal chance to be drawn. We made a difference between the colors in describing the setup. The statement, "A ball is drawn at random." tells us that each ball has an equal chance to be drawn. When a problem is posed to us, we must judge from the precise wording of the question what is intended to be equally likely.

Another question of language is best indicated by an example. Four coins are going to be tossed. To be specific, suppose they are a penny, a nickel, a dime, and a quarter. We bet that three coins will fall heads. If all four coins fall heads, do we win? We can argue, for example, that three coins, namely the penny, the nickel, and the dime, have fallen heads. Our opponent would argue that four coins fell heads and four is not three. Who is right? There is no answer because the bet was too vague. To simplify matters in the future, we adopt a convention on this point. Mathematics books differ as to what convention they choose; different areas of mathematics make different terminology convenient. In our context, the easiest way to proceed is to say that three means just three and no more. From now on, when we say something happens n times, we mean the number of times it happens is n; in other words, it happens only n times. We can still, for emphasis, make statements like, "just five of the dice fall 'six'." But the meaning would be the same if the word "just" were omitted.

Throughout the book, we shall insert various brief comments on the history of the study of probability. In particular, we shall include very short biographies of some of the people who devised probability theory. At this point, we are ready to conclude the chapter with a discussion of how it all started.

The earliest records of the study of probability theory are some letters exchanged by Blaise Pascal and Pierre Fermat in the summer of 1654. While neither man published anything about probability, and only a few particular problems were discussed in the correspondence, informal communication spread the word in the mathematical community. (A particular case of this is somewhat amusing. Fermat, in writing to Pascal, says at one point that Pascal will have already understood the solution to a certain problem. Fermat goes on to announce that he will spell out the details anyhow for the benefit of M. de Roberval. Gilles Personne de Roberval (1602–1675) was professor of mathematics at the Collége Royal in Paris; he apparently had some difficulty in understanding the work of Pascal and Fermat. In fact, Gottfried Leibniz (1646–1716) refers to "the beautiful work on chance of Fermat, Pascal, and Huygens, about which M. Roberval is neither willing nor able to understand anything." In fairness, we should at least mention that Roberval did make many important contributions in mathematics and physics.) The letter in which the topic of probability was first introduced has been lost. From the surviving letters it appears that Pascal was interested in discussing chance, while Fermat tried to persuade Pascal to turn his attention to number theory. Each of the two men suggests a way of attacking a certain type of probability problem. Fermat recommends finding probabilities by exactly the method we have suggested here: dividing the number of favorable cases by the total number of cases. Pascal is concerned with assigning monetary values to contingent events when gambling, a topic we leave for Chapter Four. Both men realized the equivalence of probability and what we shall call expected value, and, after some misunderstandings, were in complete agreement as to their conclusions.

Pierre Fermat

Pierre de Fermat (French, 1601–1665)

Fermat's father was a prosperous leather merchant. Fermat himself studied law. His father's wealth enabled Fermat to purchase a succession of judicial offices in the *parlement* of Toulouse; at that time in France, the usual way of becoming a judge was to purchase a judgeship. These positions entitled him to add "de" to his name; however he only did so in his official capacity. Despite his profession, Fermat is usually described in reference books as a mathematician. Certainly his mathematical accomplishments are enormous. Since the brief biographies, of which this is the first, will not describe the subject's mathematical accomplishments, we just conclude by pointing out that little is known about Fermat's life.

Blaise Pascal

(French, 1623–1662)

Pascal was descended from a long line of senior civil servants. In 1470 the Pascal family was "ennobled" and thus received the right to add "de" to their name; however they did not do so. Blaise Pascal's father, Etienne, had a substantial interest in mathematics. In 1637, Etienne Pascal, Fermat, and Roberval jointly sent René Descartes (1596–1650) a critical comment on his geometry. Blaise Pascal's health was very poor almost from the time of his birth; he was more or less an invalid throughout his life. Blaise, together with his two sisters, was educated at home by his father. The sisters were Gilberte, born in 1620, and Jacqueline, born in 1625. All three children were child prodigies. In 1638, Etienne protested the omission of interest payments on some government bonds he owned, and, as a result, Cardinal Richelieu (1585–1642) ordered his arrest. The family was forced to flee. Then the two sisters participated in a children's performance of a dramatic tragedy sponsored by the cardinal. After the performance, Jacqueline read a poem she had written to the cardinal, who was so impressed that he reconsidered his feelings about Etienne. As a result, Etienne was given the post of *intendent* at Rouen. This office made him the administrator of, and chief representative of the king for, the surrounding province. It is curious that Fermat was a member of a *parlement* while Etienne Pascal was an *intendent*. At this time, there was great political tension between the *intendents*, representing the central authority of the king, and the local *parlements*, representing regional interests. In fact, in a revolution that took place between 1648 and 1653, the *intendents*, including Etienne Pascal, were driven out of office. Etienne died in 1651. During the 1640s, the family had become more and more taken up with the Jansenist movement. Their ties to it had been strengthen by Gilberte's marriage in 1641. In 1652, shortly after the death of her father, Jacqueline became a nun at the Jansenist convent of Port-Royal. In 1655, Blaise entered the convent. While he turned his attention primarily to theology, he did do some important mathematical work on the cycloid in 1658. Jacqueline, broken-hearted by the pope's suppression of Jansenism, died in 1661. On March 18, 1662, an invention made by Blaise Pascal went into operation in Paris: carriages running on fixed routes at frequent intervals could be boarded by anyone on the payment of five sous. In short, Pascal invented buses. Shortly after seeing the success of his bus system, Pascal died at the age of 39. His surviving sister, who was then Gilberte Perier, was instrumental in securing the posthumous publication of much of his work.

Two years after the correspondence between Fermat and Pascal mentioned above, Christiaan Huygens became interested in probability theory. He worked out the basic ideas on his own. Then he wrote to Roberval and others to try to compare his methods with those of the French mathematicians; the methods are the same. In 1656, Huygens wrote a short work on probability in Dutch. This work was translated into Latin and published with the title, *De ratiociniis in ludo aleae*, in 1657. It thus became the first published account of probability theory. While Huygens added little of significance to the work of Fermat and Pascal, he did set out explicitly the relationship between the probability approach of Fermat and the expected-value approach of Pascal.

Christiaan Huygens

(Dutch, 1629–1695)

Christiaan Huygens's father, Constantijn, was a diplomat and poet of some note. He was also a friend of René Descartes (1596–1650). Descartes was much impressed by the young Christiaan's mathematical ability when he visited the Huygens's home. Christiaan studied both mathematics and law at the University of Leiden. He want on to devote his time to mathematics, physics, and astronomy; since he was supported by his wealthy father, he did not need to seek gainful employment. While some of his work was theoretical, he was also an experimenter and observer. He devised, made, and used clocks and telescopes. When he visited Paris for the first time in 1655, he met Roberval and other French mathematicians. In 1666, he became a founding member of the French Academy of Sciences. He was awarded a stipend larger than that of any other member and given an apartment in which to live. Accordingly he moved to Paris, remaining there even while France and Holland were at war. In 1681, serious illness forced him to return home to Holland. Various political complications made it impossible for him to go back to France, and so he spent the rest of his life in Holland.

Exercises

1. Allen, Baker, Cabot, and Dean are to speak at a dinner. They will draw lots to determine the order in which they will speak.

   a. List all the elements of a suitable sample space, if all orders need to be distinguished.

   b. Mark with a check the elements of the event, "Allen speaks before Cabot."

   c. Mark with a cross the elements of the event, "Cabot's speech comes between those of Allen and Baker."

   d. Mark with a star the elements of the event, "The four persons speak in alphabetical order."

2. An airport limousine can carry up to seven passengers. It stops to pick up people at each of two hotels.

   a. Describe a sample space that we can use in studying how many passengers get on at each hotel.

   b. Can your sample space still be used if we are concerned only with how many passengers arrive at the airport?

3. A coin is tossed five times. By counting the elements in the following events, determine the probability of each event.

   a. Heads never occurs twice in a row.

   b. Neither heads nor tails ever occurs twice in a row.

   c. Both heads and tails occur at least twice in a row.

4. Two dice are thrown. Consider the sample space with 36 elements described above. Let:

   > A be "The total is two."
   > B be "The total is seven."
   > C be "The number shown on the first die is odd."
   > D be "The number shown on the second die is odd."
   > E be "The total is odd."

   By counting the appropriate elements of the sample space if necessary, find the probability of each of the following events: a. A; b. B; c. C; d. D; e. E; f. $A \cup B$; g. $A \cap B$; h. $A \cup C$; i. $A \cap C$; j. $A \setminus C$; k. $C \setminus A$; l. $D \setminus C$; m. $B \cup \overline{D}$; n. $B \cap \overline{D}$; o. $C \cap D$; p. $D \cap E$; q. $C \cap E$; r. $C \cap D \cap E$.

5. Show that, for any events A, B, and C:

    a. $P(A \cap B) + P((A \setminus B) \cup (B \setminus A)) + P(\overline{A} \cap \overline{B}) = 1$.

    b. $P(\overline{A} \cap \overline{B}) + P(A) + P(\overline{A} \cap B) = 1$.

    c. $P(\overline{A} \cap \overline{B} \cap \overline{C}) + P(A) + P(\overline{A} \cap B) + P(\overline{A} \cap \overline{B} \cap C) = 1$.

    d. $P(A \cap B) - P(A)P(B) = P(\overline{A} \cap \overline{B}) - P(\overline{A})P(\overline{B})$.

6. Suppose $P(A) \geq .9$, $P(B) \geq .8$, and $P(A \cap B \cap C) = 0$. Show that $P(C) \leq .3$.

7. Prove: If $A \cap B \cap C = \emptyset$, then $P((A \cup B) \cap (B \cup C) \cap (C \cup A)) = P(A \cap B) + P(B \cap C) + P(C \cap A)$.

2 Counting

CHAPTER

Recall that we saw in the last chapter that determining probabilities often comes down to counting how many this-or-thats there are. In this chapter we gain practice in doing that. While occasionally we shall phrase problems in terms of probability, the solutions will always be obtained by counting, until, in later chapters, we develop more probabilistic methods.

Let us begin by working two problems. Since we are just beginning, the problems will of necessity be easy; but they serve to introduce some important basic principles. We state the first problem: Suppose we are going to appoint a committee consisting of two members of congress, one senator and one representative. In how many ways can this be done; in other words, how many different committees are possible? We may choose any one of the 100 senators. Whichever senator we choose, we may then choose any one of the 435 possible representatives to complete the committee, that is, for each choice of a senator, there are 435 possible committees. Thus in all there are $100 \cdot 435 = 43,500$ possible committees.

Now we change the conditions of the problem just worked and thereby raise a new question. We add to the original conditions the rule that the two members of the committee must be from different states. We try to proceed as before. Again we can choose any one of 100 senators. But how many representatives can serve with that senator depends on from which state the senator comes. The author does not believe he is unusual in not knowing the number of representatives from each of the states. Thus, unless we take the time to look up the information, we are stuck. The key to this whole book, especially this chapter, is, "When stuck, try another method." Let us think about choosing the representative first. We can choose any of 435 representatives. For each possible choice, two senators are excluded and 98 are available to complete the committee. Thus, for each of the 435 representatives there are 98 possible committees that include that representative. We conclude that there are $435 \cdot 98 = 42,630$ possible committees.

Now the most important part of the last two paragraphs is very general. Faced with a problem, we try to look at it from several points of view until a solution becomes obvious. Sometimes, but far from always, a formula is useful. In such cases, deciding that a formula will help, and determining which formula, is the heart of the problem. Using the formula is in itself a minor task. And, to repeat, much of the time no formula is available. That had to be the case in the problems worked so far, since we have not yet derived any formulas. It may take time to work a problem. It may appear that no progress towards a solution is being made until suddenly the answer becomes obvious. However, it *is* possible to learn to work problems that one could not have done earlier. One learns by practice, even just by trying and failing. (One does not learn much by simply following another person's reasoning. Don't expect to find many problems coming back for a second time, perhaps with a trivial change in numbers. The point is not to learn to work certain types of counting problems; the point is to learn to solve counting problems.) All that really remains to be done in this chapter is to practice solving problems.

Many of the exercises of this chapter can now be done. No special information as to how to do them will be obtained by reading the rest of the chapter. However, in other exercises it is important to have some of the thinking done in advance. In other words, in some cases a formula will be a useful tool in shortening our thinking—a formula should never replace thinking. We proceed to develop some useful formulas. We begin by recording part of the solution to our two introductory problems, about choosing a committee of members of congress, as a formula.

Suppose we are going first to select one of f alternatives and then independently to select one of s alternatives. In how many ways can we make our selection? Clearly fs ways, since each of the f possible first choices leads to s possible second choices. The last sentence contains the heart of the matter. We don't care if we have the same s possibilities for a second choice regardless of which first choice we make; what is important is the number s of possibilities. This same multiplication principle may be used repeatedly if we are concerned with more than two successive choices.

One basic type of problem runs along the following lines: We are going to make successive choices, each of one object from the same given collection of objects. Suppose there are n objects in all and we are going to make r selections, each of one of the objects. How many different ways are there to make all of the selections? The question is too vague. We don't know if the same object may be chosen more than once. If so, we speak of drawing, or choosing, "with replacement." If an object that is chosen once may never be chosen again, we say "without replacement." Another vagueness in our question is that we have not stated whether or not we distinguish two different sets of choices that differ only as to the order in which they were made. In other words, are we concerned only with which objects we have chosen when we are all done, or does the order in which we select the objects have some significance? We can say, for short, "order counts" or "order does not count." In accordance with the multiplication principle of the last paragraph, we have four problems to do. In increasing order of difficulty they are:

1. order counts, with replacement;

2. order counts, without replacement;

3. order does not count, without replacement;

4. order does not count, with replacement.

We shall now solve these four problems, with some digressions along the way.

1.   order counts, with replacement

For each of the r selections from the n objects, we can choose any one of the objects, independently of how we make the other selections. Repeated application of the multiplication principle stated above tells us that there are

$$n^r$$

ways to make the selections in this case.

We work an example using this last formula. How many six-letter computer words can be formed using the letters A, B, C, D, E, with repetitions permitted. We say "computer words" to indicate that we allow any six letters in a row; BBBCCC counts just as well as ACCEDE, even though the former is not an English word. (In other problems, we will even use characters, such as digits, that are not letters.) In the case at hand, we must choose which letter to put in each of six different positions; thus $5^6 = 15{,}625$ answers our question.

2.   order counts, without replacement

In the first place, since no object may be chosen more than once, the choice cannot be made at all if $r > n$. We consider in turn the values $1, 2, 3, \ldots, n$ for r. If only one object is to be chosen, this one object can be any one of the n objects; thus there are n ways to make the selection. If two objects are to be chosen, the first can be chosen in n ways, as we just said, and each of these ways gives us $n - 1$ possible choices for the second object; thus there are $n(n - 1)$ ways to select the two objects. If three objects are to be chosen, we just saw that there are $n(n - 1)$ ways to choose the first two objects, and each of these ways leads to $n - 2$ possibilities for the third object; thus there are $n(n-1)(n-2)$ ways to make all three choices. Continuing in this manner, we obtain the answer we desired: For $r \le n$, there are

$$n(n - 1) \cdots (n - r + 1)$$

ways to make the selection under consideration. (By convention, this notation means the values already explicitly stated in case $r = 1$, 2, or 3.)

The notation $n(n-1) \cdots (n-r+1)$ just obtained is a little awkward to work with; it is sometimes more convenient to express the number in a different way. If we multiply $n(n-1) \cdots (n-r+1)$ by $(n-r)! = (n-r)(n-r-1) \cdots 1$, the product is $n!$. Thus, $n(n-1) \cdots (n-r+1) = n!/(n-r)!$. The form $n!/(n-r)!$ will be especially useful after we obtain, in Chapter Six, a method of approximating $k!$ when k is so large that the direct computation of $k!$ is difficult, or even impossible.

We give an example of the use of the formula just derived. Suppose five of a group of 100 people are each to receive one of five prizes, each prize having a distinctive name. Disregarding the order in which the prize recipients are chosen, but taking into account that the prizes are all different, in how many ways can the five recipients be selected from the 100 candidates? The first step is typical of many problems along these lines. If objects are being regarded as different, they may as well be regarded as arranged in order. Specifically, consider the prizes. We are not told their names, but the names might be "First Prize," "Second Prize," etc. It is thus clear that the problem of choosing *in order* five people to receive five identical prizes is equivalent to the given problem of choosing five people to receive five different prizes. Thus our problem as originally stated amounts to choosing five out of 100 people, order counts, without replacement. The answer is thus $100!/95! = 9{,}034{,}502{,}400$.

An important special case is $n = r$. Then all the objects are selected, and we are concerned only with the order in which they are selected. In other words, we are finding the number of ways of arranging n objects in order. With the standard convention $0! = 1$, the formula $n!/(n-r)!$ just obtained gives $n!$ as the number of ways that n objects can be arranged in order so that there is a first, a second, etc.

We just found the number of ways we can arrange n objects in a row. Sometimes we want to arrange them in other ways, such as in a circle. It can be helpful to think of the places the objects will occupy as preexisting. Then careful application of the formula just derived can be helpful. The special case of arranging n objects in a circle is the same problem as that of seating n people at a round table. We next give an example of just that.

The president and ten members of the cabinet are to sit at a round table. In how many ways can they be seated? To make the question interesting, we adopt a certain natural convention. The whole point of a round table is that all seats are alike. If each person moves one seat to the right, the arrangement of the people is still the same. As between different arrangements, someone would have to have a different right-hand neighbor. When we consider a round table, we want to determine how many different, in that sense, arrangements are possible.

Now we return to the president and cabinet. As is often the case, introducing a time factor will help explain our computation. Of course, the president sits down first. It makes no difference where the first person to be seated sits; all chairs at a round table are alike. After the president is seated, the position of each chair may be specified in relation to the president's chair. For example, one chair is four chairs to the president's right; some chair is three chairs to the president's left, etc. Thus after the president is seated, we consider the number of ways ten people can be placed into ten different chairs. This can be done in $10! = 3,628,800$ ways, and that is the answer to our question.

It is now a trivial exercise to give a formula for the number of ways n people can sit at a round table. Problems involving other shapes of tables can be done on the basis of analogous reasoning.

The next topic, which is also a digression, is most easily described by way of examples. In how many ways can the letters of the word SCHOOL be arranged? There are some obvious conventions in understanding what the question means. The letters are to be arranged in a row from left to right, as we normally write letters. Thus HOOLCS is one arrangement; SCLOOH is another. The point of the question is that the two Os cannot be distinguished from each other. The way to answer the question is first to compute the number of arrangements there would be if the Os differed, say by being printed in different fonts, and then adjusting. Six distinct letters can be arranged in $6! = 720$ ways. But in fact we cannot tell the two Os apart. Thus the 720 ways counts SHOL*O*C and SH*O*LOC as different, when they should have been counted as the same. Clearly the answer to our question is 720 divided by 2, since either O can come first. In other words, the answer is $720/2 = 360$, because the Os can be arranged in order in two ways.

Now let us try a more complicated problem. In how many ways can the letters of the phrase

SASKATOON SASKATCHEWAN

be arranged? Note that the question says "letters"; the space is to be ignored. The first step is simply to count how many times each distinct letter appears. There are five As; four Ss; two each of K, T, O, and N; and one each of C, H, E, and W. That makes a total of 21 letters, counting repetitions. If all the letters were different, there would be 21! ways they could be arranged. As in the last problem, we divide by two because there are two Os. Then we again divide by two for the two Ks; then again for the Ts; and finally a fourth time for the Ns. At this point we have $21!/2^4$. There are four Ss; they can be arranged in 4! ways. Thus there are 4! times as many arrangements if we distinguish the Ss as there are if we do not. We divide by 4! because we do not want to distinguish the Ss. Finally we divide by 5! because the five As are all alike. Our final answer is thus

$$\frac{21!}{5!\,4!\,(2!)^4}.$$

To make the procedure by which we obtained this number clearer, we may write the number as

$$\frac{21!}{5!\,1!\,1!\,1!\,2!\,2!\,2!\,4!\,2!\,1!},$$

where the numbers 5, 1, 1, 1, 2, 2, 2, 4, 2, 1 refer to the letters A, C, E, H, K, N, O, S, T, W in alphabetical order. (With a calculator, we find that the answer just given is roughly eleven hundred trillion.) Since the ideas here are not made any clearer by writing out a general formula, we shall just assume that the method is now obvious and go on.

3. order does not count, without replacement

The idea here is to build on the last item we discussed, item 2. In both items 2 and 3, we are selecting r objects out of n, without replacement. In both, we must have $r \le n$ to avoid the answer zero. We saw that there were $n!/(n-r)!$ ways to make the selection if order does count; denote this number by y. We are now trying to find the number of ways to make the selection when order is not counted; denote this number by x. We may think of choosing the objects with order counting as a two-stage process. First decide which objects to use, and then arrange them in order. Each choice of a certain r object leads to $r!$ arrangements of these objects in order. Thus $xr! = y$. Obviously then,

$$x = \frac{y}{r!} = \frac{n!}{(n-r)!} \cdot \frac{1}{r!} = \frac{n!}{r!(n-r)!}.$$

We use the notation $\binom{n}{r}$ for this number; thus

$$\binom{n}{r} = \frac{n!}{r!\,(n-r)!}$$

is the number of ways of choosing r objects out of n, without replacement, disregarding the order of selection.

We illustrate the use of this formula. How many different committees of 10 United States senators can be formed? Since we are choosing 10 out of 100 senators, the answer is

$$\binom{100}{10} = \frac{100!}{10!\,90!} = \frac{100 \cdot 99 \cdot 98 \cdot 97 \cdot 96 \cdot 95 \cdot 94 \cdot 93 \cdot 92 \cdot 91}{10 \cdot 9 \cdot 8 \cdot 7 \cdot 6 \cdot 5 \cdot 4 \cdot 3 \cdot 2 \cdot 1}$$

which comes to roughly 17 trillion.

It is instructive, and useful from the point of view of generalization, to also give an alternative derivation of the last formula obtained. Suppose we begin by placing all n objects in a row. Then we make our selection of r of the objects by marking these r with the letter A. Then we mark the other $n - r$ objects B. The result is that we now have arranged in a row n letters; r of these letters are As and $n - r$ of them are Bs. Each way of choosing r objects corresponds to an arrangement of the letters, and vice versa. Thus there are just as many ways to pick the objects as there are ways to arrange the letters. We already know that there are

$$\frac{n!}{r!\,(n-r)!}$$

ways to arrange the letters. Thus we have a new proof of the formula of the last paragraph.

Let us take a moment to generalize the argument just concluded. Choosing r objects out of n, without replacement, amounts to dividing the objects into two categories—those that are chosen and those that are not. We designated these categories with A and B, respectively. In exactly the same way we can consider dividing the objects into more than two categories. We do retain the condition that the number of objects in each category be determined in advance. Suppose that there are k categories, and that we plan to put r_1 objects into the first category, r_2 into the second, and so on. If we can do this at all, we must have $r_1 + r_2 + \cdots + r_k = n$, since each object is placed in just one category. We reason as in the last paragraph. Put all n objects in a row. We make our selection by marking the objects put in to the first category with the letter A, those in the second category with B, those in the third with C, and so on. The number of ways to divide the objects into categories is the same as the number of ways to arrange a certain collection of n letters, not all distinguishable, in a row. We saw earlier that this number is

$$\frac{n!}{r_1! r_2! \cdots r_k!}$$

4.   order does not count, with replacement

Curiously enough, after we solve the practical problem of keeping track of which objects are selected how many times, the theoretical problem of determining the number of ways in which the selection can be made becomes easy. The practical difficulty in knowing how many times each object has been chosen is that we must replace each object as soon as we choose it. Obviously, one would overcome this difficulty by keeping a written record. One way to do this is the following: Divide a sheet of paper into n compartments by drawing vertical bars. The diagrams illustrate the example, $n = 5$, $r = 12$. In that example, we divide our paper into five compartments by drawing four vertical bars, thus:

$$|\quad|\quad|\quad|$$

In general, the number of bars is $n - 1$, one less than the number of compartments. We assign a different one of the n compartments to each of the n objects. Now we start choosing objects. Each time an object is selected, we put a star in the compartment corresponding to that object. We may as well put all the stars in one horizontal row. Then, when we have made all the selections, we have stars and bars arranged in a row. (Only the relative order of the stars and bars counts; the amount of space between them has no significance.) For example, we could have:

$$**|****||***|***$$

In this example, the first object was selected twice, the second object four times, the third object was never selected, and the fourth and fifth objects were selected three times each. Each way of making r selections from n objects, with replacement, disregarding order of selection, corresponds to just one arrangement of the stars and bars. Furthermore, each such arrangement corresponds to a certain selection. Thus the number we seek is simply the number of ways to put the stars and bars in order. As noted above, there are $n-1$ bars. Since there is a star for each selection, there are r stars. The number of ways of arranging $n-1$ bars and r stars in a row was discussed earlier; it is

$$\frac{(n-1+r)!}{(n-1)!\,r!}.$$

We now have the notation

$$\binom{n-r+1}{r}$$

for this number.

    We give a couple of applications of this last formula. In how many ways can 14 chocolate bars be distributed among five children? For each of the 14 bars we choose one of the five children to receive it. Thus, in our formula, $n=5$, $r=14$ and the answer is

$$\binom{5+14-1}{14} = \binom{18}{14} = 3060.$$

Suppose, in the interest of greater fairness, we require each child to receive at least two of the chocolate bars. Since the bars are alike, we may as well first distribute the ten bars necessary to meet the guaranteed minimum of two bars per child. The number of ways the remaining four bars can be distributed is given using $n = 5$, $r = 4$ in our formula; thus this time our answer is

$$\binom{5+4-1}{4} = \binom{8}{4} = 70.$$

We have been discussing choosing r objects out of n objects. Thus we are, of course, regarding n and r as positive integers. Occasionally it is convenient to have a definition for the symbol $\binom{n}{r}$ with $r = 0$. The formula

$$\binom{n}{r} = \frac{n!}{r!\,(n-r)!},$$

together with the usual definition $0! = 1$, gives

$$\binom{n}{0} = 1$$

for all nonnegative integers n. We take this last equation as the definition of the symbol $\binom{n}{r}$ for $r = 0$, with n a nonnegative integer.

We digress briefly to describe an important application of the numbers just considered. Consider the familiar problem of evaluating

$$(A + B)^n.$$

By multiplying out, or, to say the same thing in more formal language, by applying the distributive laws, *without* simplifying at all, we find that $(A+B)^n$ is the sum of many terms. Each term is obtained by choosing either the A or the B from each of the n factors in

$$(A + B)(A + B) \cdots (A + B)$$

and listing the chosen letters in order. Now we simplify. First each term can be written as

$$A^r B^{n-r},$$

where r is an integer with $0 \le r \le n$. For a given r, the number of repetitions of $A^r B^{n-r}$ that appear is the number of ways to arrange r letter As and $n - r$ letter Bs in a row; we recall that this number is

$$\frac{n!}{r!\,(n-r)!}.$$

The total of all terms with a given r is thus

$$\binom{n}{r} A^r B^{n-r}.$$

Summing over all r, we have

$$(A+B)^n = \binom{n}{0} A^n + \binom{n}{1} A^{n-1}B + \binom{n}{2} A^{n-2}B^2 + \cdots + \binom{n}{n} B^n.$$

This last equation is the well-known *Binomial Theorem*. Because of their appearance in this theorem, the numbers

$$\binom{n}{r}$$

are called *binomial coefficients*.

In a corresponding way, we could develop the *Multinomial Theorem* that replaces $A + B$ with the sum of more than two letters. The numbers

$$\frac{n!}{r_1! r_2! \cdots r_k!},$$

which we mentioned earlier, would appear in that theorem and are therefore called *multinomial coefficients*.

As pointed out above, choosing k objects out of n objects, disregarding order, without replacement, amounts to dividing the n objects into two categories—those that are selected and those that are rejected. In other words, selecting k objects amounts to rejecting $n - k$ objects. There are $\binom{n}{n-k}$ possibilities for which $n - k$ objects are rejected. Reasoning that way, or using the formula

$$\binom{n}{r} = \frac{n!}{r!\,(n-r)!},$$

we see that

$$\binom{n}{k} = \binom{n}{n-k},$$

whenever n and k are integers with $0 \le k \le n$.

Next we shall derive a formula that is very useful in finding $\binom{n}{r}$ for small values of n. For larger values of n, the formula is still useful because it is helpful in discussing theory. We suppose neither n nor r is zero; we also suppose $r \neq n$. Suppose we are going to choose r objects out of n, without replacement, where the order of selection does not count. There are $\binom{n}{r}$ ways to choose the objects. Let us classify these ways according to whether a certain object is chosen or not. To make the wording easier, we may assume that just one of the objects is red; judicious use of paint can always bring that about. In choosing r objects from the n, we may or may not include the red object. If the red object is among those that are selected, we also have to select $r - 1$ other objects. These $r - 1$ other objects are selected from the $n - 1$ objects that are not red. Thus in choosing r objects out of our n objects, there are $\binom{n-1}{r-1}$ ways to make the selection if the red object is to be included. Now turn to the case where the red object is not picked. Then we must make all our r choices from the $n - 1$ objects that are not red. We can do this in $\binom{n-1}{r}$ ways. To summarize, of the $\binom{n}{r}$ ways to pick r objects from n objects, just one of which is red, $\binom{n-1}{r-1}$ involve picking the red object and $\binom{n-1}{r}$ involve leaving the red object behind. Since the red object either is chosen or it is not,

$$\binom{n}{r} = \binom{n-1}{r-1} + \binom{n-1}{r}.$$

The formula derived in the last paragraph may be used to draw up a table of binomial coefficients $\binom{n}{r}$. We use a row of the table for each value of n, beginning at the top with $n = 0$. In each row, we list the coefficients for $r = 0, 1, \ldots, n$ in that order. For convenience in using the formula of the last paragraph, we begin each row a little further to the left than the row above, so that $\binom{n}{r}$ appears below and to the left of $\binom{n-1}{r}$, but below and to the right of $\binom{n-1}{r-1}$. [Of course, the last sentence tacitly assumes that $\binom{n-1}{r}$ and $\binom{n-1}{r-1}$ are defined; that is, we need $n \neq 0$, $r \neq 0$, and $r \neq n$.] We begin the table with $\binom{0}{0} = 1$; 1 is the only entry in the top row of the table. The next row contains $\binom{1}{0}$ and $\binom{1}{1}$, both of which are 1. For esthetic reasons, we insert these 1s as follows:

$$1$$
$$1 \qquad 1$$

The next row lists $\binom{2}{0}$, $\binom{2}{1}$, $\binom{2}{2}$, namely, 1, 2, 1. As stated above, these numbers are placed as follows:

$$
\begin{array}{ccccc}
 & & 1 & & \\
 & 1 & & 1 & \\
1 & & 2 & & 1
\end{array}
$$

Now we consider $\binom{n}{r}$ with $n = 3$. $\binom{3}{0} = \binom{3}{3} = 1$ are the first and last entries in the row. The others are

$$
\binom{3}{1} = \binom{2}{0} + \binom{2}{1} = 1 + 2 = 3,
$$

$$
\binom{3}{2} = \binom{2}{1} + \binom{2}{2} = 2 + 1 = 3.
$$

We show this computation by

$$
\begin{array}{ccccccc}
 & & & 1 & & & \\
 & & 1 & & 1 & & \\
 & 1 & & 2 & & 1 & \\
1 & & 3 & & 3 & & 1
\end{array}
$$

We continue in this way. Each row begins and ends with a 1. Each of the other entries, according to the formula

$$
\binom{n}{r} = \binom{n-1}{r-1} + \binom{n-1}{r},
$$

is the sum of the two numbers most nearly directly above it. The array described above, part of which is shown in the table below, is known as *Pascal's Triangle*. (A short biography of Pascal appears in Chapter One.) However the triangle was invented by Jia Xian, sometimes spelled Chia Hsien, in China in the eleventh century. Jia Xian devised his triangle to use in raising sums of two terms to powers. Even in Europe, something was known about the triangle before the time of Pascal. But the important properties of the triangle were first formally established by Pascal. We note in passing that Pascal invented the method now called "mathematical induction" for just this purpose. Pascal's Triangle provides a convenient way to compute binomial coefficients when we need all $\binom{n}{r}$ for a given, not too large, n. The triangle begins:

```
                              1
                           1     1
                        1     2     1
                     1     3     3     1
                  1     4     6     4     1
               1     5    10    10     5     1
            1     6    15    20    15     6     1
         1     7    21    35    35    21     7     1
      1     8    28    56    70    56    28     8     1
   1     9    36    84   126   126    84    36     9     1
 1    10    45   120   210   252   210   120    45    10     1
1   11    55   165   330   462   462   330   165    55    11     1
1   12    66   220   495   792   924   792   495   220    66    12     1
```

Exercises

1. a. In how many ways can the letters of the word

   FLUFF

   be arranged?

   b. In how many ways can the letters of the word

   ROTOR

   be arranged leaving the T in the middle?

   c. In how many ways can the letters of the word

   REDIVIDER

   be arranged so that we still have a palindrome, that is, the letters read the same backwards as forwards?

2. a. In how many ways can the letters of the word

   SINGULAR

   be arranged?

   b. In how many ways can the letters of the word

   DOUBLED

be arranged?

c. In how many ways can the letters of the word

REPETITIOUS

be arranged?

3. a. In how many ways can the letters of the word

KRAKATOA

be arranged?

b. In how many ways can the letters of the word

MISSISSIPPI

be arranged?

c. In how many ways can the letters of the phrase

MINNEAPOLIS MINNESOTA

be arranged?

d. In how many ways can the letters of the phrase

NINETEEN TENNIS NETS

be arranged?

4. Suppose your campus bookstore has left in stock three copies of the calculus book, four copies of the linear algebra book, and five copies of the discrete probability book. In how many different orders can these books be arranged on the shelf?

5. How many numbers can be made each using all the digits 1, 2, 3, 4, 4, 5, 5, 5?

6. How many numbers can be made each using all the digits 1, 2, 2, 3, 3, 3, 0?

7. Five persons, A, B, C, D, and E, are going to speak at a meeting.

a. In how many orders can they take their turns if B must speak after A?

b. How many if B must speak immediately after A?

8. In how many ways can the letters of the word

MUHAMMADAN

be arranged without letting three letters that are alike come together?

9. At a table in a restaurant, six people ordered roast beef, three ordered turkey, two ordered pork chops, and one ordered flounder. Of course, no two portions of any of these items are absolutely identical. The 12 servings are brought from the kitchen.

   a. In how many ways can they be distributed so that everyone gets the correct item?

   b. In how many ways can they be distributed so that no one gets the correct item?

10. a. In how many ways can an American, a Dane, an Egyptian, a Russian, and a Swede sit at a round table?

   b. In how many ways can an amethyst, a diamond, an emerald, a ruby, and a sapphire be arranged on a gold necklace?

11. In how many ways can five men and five women sit at a round table so that no two men sit next to each other?

12. In how many ways can five men and eight women sit at a round table if the men sit in consecutive seats?

13. In how many ways can eight people, including Smith and Jones, sit at a round table with Smith next to Jones?

14. Mr. Allen, Mrs. Allen, Mr. Baker, Mrs. Baker, Mr. Carter, Miss Davis, and Mr. Evans are to be seated at a round table. What is the probability each wife sits next to her husband?

15. a. Twenty-two individually carved merry-go-round horses are to be arranged in two concentric rings on a merry-go-round. In how many ways can this be done, if each ring is to contain 11 horses and each horse is to be abreast of another horse?

   b. Suppose instead that 12 of the horses are to be placed on one single-ring merry-go-round and the other ten on another single-ring merry-go-round?

16. a. In how many ways can eight people sit at a lunch counter with eight stools?

   b. In how many ways can eight people sit at a round table?

   c. In how many ways can four couples sit at the lunch counter if each wife sits next to her husband?

    d. In how many ways can four couples sit at the round table if each husband sits next to his wife?

    e. In how many ways can four couples sit at a square table with one couple on each side?

17. At a rectangular dining table, the host and hostess sit one at each end. In how many ways can each of the following sit at the table:

    a. six guests, three on each side?

    b. four male and four female guests, four on each side, in such a way that no two persons of the same sex sit next to each other?

    c. eight guests, four one each side, so that two particular guests sit next to each other?

**18. Show that four persons of each of n nationalities can stand in a row in $12^n(2n)!$ ways with each person standing next to a compatriot.

19. A store has in stock one copy of a certain book, two copies of another book, and four copies of a third book. How many different purchases can a customer make of these books? (A purchase can be anything from one copy of one book to the entire stock of the store.)

20. A restaurant offers its patrons the following choices for a complete dinner:

    i. choose one appetizer out of four;

    ii. choose one entree out of five;

    iii. choose two different items from a list of three kinds of potatoes, three vegetables, and one salad;

    iv. choose one dessert out of four;

    v. choose one beverage out of three.

    a. How many different dinners can be ordered without ordering more than one kind of potato, assuming that no course is omitted?

    b. How many different dinners can be ordered with no more than one kind of potato if one item, other than the entree, is omitted?

21. Of 12 men, just two are named Smith. In how many ways, disregarding the order of selection, can seven of the men be chosen: a) with no restrictions? b) if both Smiths must be included? c) if neither Smith may be included? d) if just one Smith must be included? e) if at least one Smith must be included? f) if no more than one Smith may be included?

22. Consider computer words consisting of six characters each. If the characters are chosen from a, b, c, d, 2, 3, 4, 5, 6, each of which may be used at most once, how many words can be formed: a) with no other restrictions? b) if the third and fourth characters must be digits and the other characters must be letters? c) If there must be three digits and three letters? d) if no two digits and no two letters may be adjacent?

23. Do the last problem with the modification that the characters may be used more than once.

24. For each of $k = 0, 1, 2, 3, 4$, find the probability that a poker hand (five cards) contains just k aces.

25. Of ten twenty-dollar bills, two are counterfeit. Six bills are chosen at random. What is the probability that both counterfeit bills are chosen?

26. An urn contains eight balls—two red, two blue, two orange, and two green. The balls are separated at random into two sets of four balls each. What is the probability that each set contains one ball of each color?

27. A bin contains 100 balls numbered consecutively from 1 to 100. Two balls are chosen at random without replacement. What is the probability that the total of the numbers on the chosen balls is even?

28. What is the probability that a bridge hand (13 cards) contains all four aces?

29. a. How many five-person committees can be chosen from a club with 15 members?

    b. In how many ways can a five-person committee consisting of a chair, a secretary, and three other members be chosen from the club?

30. a. In how many ways can a club with 15 members be divided into three committees of five members each, with no member serving on more than one committee, if each committee has different duties?

    b. If all committees perform the same functions?

    c. In how many ways can the club be divided into three committees, each consisting of a chair, a secretary, and three other members, if the committees have different duties and no one serves on more than one committee?

31. A student committee consists of five women and four men. It is to be divided, by chance, into two working groups, one group having three members and one group having six members. John and Joan are on the committee and hope to work together. What is the probability that they will not be disappointed?

32. A poker hand contains five cards. Find the probability that a poker hand be:

    a. "four of a kind" (contains four cards of equal face value);

    b. "full house" (three cards of equal face value and two others of equal face value);

    c. "three of a kind" (three cards of equal face value and two cards with face values different from the three and from each other);

    d. "one pair" (two cards of equal face value and three cards with face values different from the two and from each other);

    e. "two pairs" (two cards of equal face value, two other cards of equal face value different from the value of the first two, and one card with face value different from each of the other four).

33. A class of 30 children comes into a store that sells ice cream cones. There are five flavors available. Each child gets just one cone. The management of the store does not care who gets which flavor. From the point of view of the management, how many different selections of flavors are possible?

34. In how many ways can 22 identical bottles of soda be distributed among four people, disregarding the order of distribution, so that each person gets at least one bottle?

35. In how many ways can five apples and six oranges be distributed among seven children, disregarding the order of distribution?

36. How many ways are there to choose three letters from the phrase

MISS MISSISSIPPI NEVER EVER SIMPERS,

ignoring the order of selection.

37. How many different computer words of six characters each can be made with letters chosen from the phrase

E PLURIBUS UNUM?

38. A store has in stock 20 cans of each of four flavors of a particular brand of soda.
    a. In how many ways can a customer purchase ten of these cans of soda?
    b. In how many ways can a customer purchase 65 of these cans of soda?

39. What is the probability that four randomly chosen people were born on four different days of the week?

40. A die is thrown four times. What is the probability that each number thrown is higher than all those that were thrown earlier?

41. A die is thrown four times. What is the probability that each number thrown is at least as high as all of the numbers that were thrown earlier?

42. In how many ways can 11 # signs and 8 * signs be arranged in a row so that no two * signs come together?

43. In how many ways can 11 men and 8 women stand in a row so that no two women stand next to each other?

**44. In how many ways can 11 women and 8 men sit at a round table so that no two men sit next to each other?

45. In how many orders can the letters of the word

INDIVISIBILITY

be arranged without having two Is together.

46. One arrangement of the letters of the word

MISSISSIPPI

is chosen at random.

a. What is the probability no two Ss are together?

b. What is the probability all the Ss are together?

47. How many arrangements of the letters of the phrase

MINNEAPOLIS MINNESOTA

have no two vowels together?

48. In a certain city the streets all run north–south and the avenues run east–west. The avenues are numbered consecutively, and the streets are denoted by letters in alphabetical order. A police car is patrolling the city. At each intersection it either goes straight ahead, makes a left turn, or makes a right turn—it never makes a U-turn. The car starts by entering the intersection of J Street and Twelfth Avenue coming from I Street. How many different routes can it take if:

a. it reaches L Street and Eighth Avenue after going just six blocks?

b. it returns to J Street and Twelfth Avenue after going four blocks?

c. it returns to J Street and Twelfth avenue after going six blocks?

49. In the city described in the last problem, suppose a police car is at B Street and Second Avenue when it gets word of an accident at J Street and Tenth Avenue. If the car is instructed to proceed to the accident by as short a route as possible, from how many routes can the driver choose?

50. How many collections of letters, in no particular order, can be formed by choosing one or more letters from the phrase

MISSISSIPPI RIVER?

**51. How many computer words of six characters each can be made by choosing letters from the phrase

NINETEEN TENNIS NETS?

(Consider various cases separately.)

52. a. A disgruntled letter carrier has to place ten different letters into seven mail boxes. He pays no attention to the addresses

on the letters. In how many ways can he distribute the letters into the boxes?

   b. Suppose he has ten identical circulars, instead of the letters, to place. In how many ways can he distribute the circulars?

53. In the New York State "Numbers" game, three digits are drawn, with replacement, each evening on television. The player chooses, in advance of course, three digits and may pick from several types of bets:

   a. "Straight": To win the player's digits must match those drawn in the order they were drawn.

   b. "Six-Way Box": The player chooses three different digits and wins if they match those drawn in any order.

   c. "Three-Way Box": The player announces a digit once and a different digit twice. To win, these three digits must match the three digits drawn in any order.

Find the probability of winning each of these bets.

54. In the New York State "Win 4" game, four digits are randomly drawn, with replacement. The player chooses four digits and may pick from several types of bets:

   a. "Straight": To win, the player's digits must match those drawn in the order they were drawn.

   b. There are four kinds of "box" bets. In each, the player wins by matching the digits drawn in any order. "24-Way Box"· The player's digits are all different. "12-Way Box": The player names one digit twice. "6-Way Box": The player names two different digits twice each . "4-Way Box": The player names one digit three times and a different digit once. Find the probability of winning each of these bets.

55. New York State's "Lotto" game is played with 54 balls, numbered from 1 to 54. The player picks six of these numbers. The next Wednesday or Saturday night, the balls are placed in the Lotto draw machine. The machine automatically releases six balls, at random; the numbers on these six balls are called the winning numbers. Then the machine releases a seventh ball; the number on this ball is called the supplementary number. There are several ways to win:

a. To win the Jackpot, the player must have picked all six winning numbers (in any order).

b. To win a Second Prize, the player must have picked five winning numbers.

c. To win a Third Prize, the player must have picked four winning numbers.

d. To win a Fourth Prize, the player must have picked three winning numbers and the supplementary number.

Find the probability of winning each of the prizes.

56. According to Laplace, in his time the lottery of France was played as follows: Ninety numbers were used and five were randomly drawn, without replacement, each time the game was played. The player could make a bet on from one to five different numbers and won if *all* these numbers were drawn. (Of course, the amount won depended on how many numbers the player chose.) Find the probability that the player won for each of the five possible kinds of bet.

**57. Let

$$m = \binom{n}{2}.$$

Show that

$$\binom{m}{2} = 3\binom{n+1}{4}$$

**58. Show that

$$\binom{n}{k} \quad \text{and} \quad \binom{2n}{2k}$$

have the same parity, that is, either both are odd or both are even.

**59. Show that there are infinitely many rows of Pascal's Triangle that consist entirely of odd numbers.

**60. We have m balls of each of three colors. We also have three boxes, each of a different shape. In how many ways can the balls be distributed among the boxes so that each box contains m balls?

3

Independence and Conditional Probability

In this chapter we study how the knowledge that one event definitely occurs influences our judgement as to the chances for some other event to occur. The case we consider first is that in which there is no influence.

3.1 Independence

We begin by describing the intuitive background. We may as well work with specific numbers. Suppose $P(A) = 1/3$ and $P(B) = 1/4$. Then, in many repetitions of our experiment, A occurs one-third and B occurs one-fourth of the time. If the occurrence of A does not change the likelihood of B, then B would occur on one-fourth of those occasions when A occurs. Since A occurs one-third of the time, $A \cap B$ would occur one-fourth of one-third of the time. In symbols, $P(A \cap B) = (1/3)(1/4) = 1/12$. Of course, the conclusion $P(A \cap B) = P(A)P(B)$ would remain valid if we used different numbers.

Now, with the last paragraph in the back of our minds, we make the following definition: Two events A and B are called *independent* if $P(A \cap B) = P(A)P(B)$. To repeat, the statement that A and B are independent simply means that the equation just given holds. (Experience leads the author to include a warning. "Independent" and "mutually exclusive" are different terms with very different meanings. In fact, the terms are almost contradictory; see Exercise 1.)

How do we know that two events are independent? One way is, of course, to verify the equation that defines independence. More often, independence is simply part of the definition of the problem we are considering. For example, suppose a pair of dice is thrown twice. Let A be "The first throw results in an odd total." Let B be "The second throw gives a total of seven or more." Then A and B are independent on the basis of common sense; how the first throw comes out does not affect the outcome of the second throw in any way. What we are really saying is that we understand the description of the experiment to imply that the events are independent. Whenever we have this kind of physical independence, we shall assume that we have mathematical independence. Our intuitive understanding of the real world suggests that that is in fact the case. As a matter of abstract mathematics, to repeat, all we can do is to regard the independence as part of the definition of the problem.

Most of the cases of independent events that arise in our discussion of probability are the kind of physical independence just mentioned. Occasionally, however, we may have a different situation. We illustrate this alternative possibility with an example. The king of spades, the queen of hearts, the three of diamonds, and the two of clubs are placed in a hat. One card is drawn from the hat at random. Note that to know the face value of the chosen card is to know the suit, and vice versa. Nevertheless, the following is true. Let A be the event, "The chosen card is black." Let B be the event, "The chosen card is a picture card." Clearly both $P(A)$ and $P(B)$ are $1/2$. $A \cap B$ occurs just when the king of spades is drawn. Thus, $P(A \cap B) = 1/4$. Since $P(A \cap B) = P(A)P(B)$, we see that A and B are independent. The point is that knowing that a black card is drawn gives us no indication of how likely the card is to be a picture card.

Now that we have the concept of independence, we are able to consider the following rather curious example. A man says his name is John Doe and offers to make a bet with you. He begins to describe the bet with a demonstration. John selects some cards from the deck and arranges them into three stacks. He lets you see which cards are in each stack. You note that:

Stack S contains 2 tens and 3 threes.
Stack T contains 1 nine, 3 sevens, 3 sixes, and 3 fives.
Stack U contains 4 eights and 2 twos.

John offers to give you first choice of stack; you may select whichever stack you think is best. Then he will choose from the remaining stacks the one he wants. Next you will pick a card at random from your stack and John will pick a card at random from his. High card wins. Your opponent is willing to bet at even money despite the fact that you get first choice of stack. Is the bet to your advantage?

Let us compute your chances of winning. Since which stacks are used is not a matter of chance, but of choice, we must make several separate computations.

First, suppose you select stack U. Then John is free to choose stack S; suppose he does so. If you draw a two, you lose. In order for you to win, you must drawn an eight and your opponent must draw a three. The probabilities of these two independent events are 2/3 and 3/5. Thus the probability that you win is only $(2/3)(3/5) = 2/5$; more likely than not, you will lose.

Perhaps you think you'll do better by using stack S. Suppose you do choose S; then John can select T. With these choices, you will win if you draw a ten and lose with a three, regardless of which card your opponent picks from T. Thus the probability that you win is again 2/5 and you must try to do better.

By now you may think T is the stack to choose. Suppose you pick stack T and your opponent then picks stack U. He will win if he gets an eight while you get a seven, six or five; otherwise, you win. The probability that he gets an eight is 2/3; the probability that you will get a seven, six or five is 9/10. Thus the probability that that you win is $1 - (2/3)(9/10) = 2/5$.

In short, if you bet at all, you will probably lose. Accordingly, you shouldn't bet. Later on, in Chapter Five, we will see that, assuming that you are fool enough to play many times, you are essentially sure to come out behind. How long it takes for you to go broke will be discussed in Chapter Eight.

Let us consider the problem of when we should call more than two events independent. We begin with three events, A, B, and C. We would not call the events independent unless

$$P(A \cap B \cap C) = P(A)P(B)P(C).$$

But that equation by itself is not enough. For example, if $C = \emptyset$, the equation holds no matter what A and B are. If A, B, and C deserve to be called independent, then surely A and B should be independent. On the other hand, we have the following example: Suppose two dice are thrown. Let A be "The first die falls odd." Let B be "The second die falls odd." Let C be "The total on the dice is odd." Clearly A and B are independent. Direct computation verifies that $P(A \cap C) = P(A)P(C)$; thus, A and C are independent. Likewise, B and C are independent. Should we say that A, B, and C are independent? No; when A and B both happen, C cannot happen. Looked at another way, $P(A \cap B \cap C) = 0$, while $P(A)P(B)P(C) = (1/2)(1/2)(1/2) = 1/8$; thus A, B, and C should not be called independent.

Our conclusion is the following definition: A, B, and C are called independent if all four of the following equations hold:

$$P(A \cap B \cap C) = P(A)P(B)P(C),$$
$$P(A \cap B) = P(A)P(B),$$
$$P(A \cap C) = P(A)P(C),$$
$$P(B \cap C) = P(B)P(C).$$

More generally, we say that the events in a certain collection of events are independent when the following condition holds: Whenever $A_1, A_2, \ldots, A_n$ are distinct events chosen from the collection, then

$$P(A_1 \cap A_2 \cap \cdots \cap A_n) = P(A_1)P(A_2) \cdots P(A_n).$$

Note that "whenever" means that we have many statements to verify; we must check the equation for each choice of $A_1, \ldots, A_n$. If a single one of the equations fails to hold, the events are not independent. However, in many cases, all the equations are obvious. For example, if the events $B_1, B_2, \ldots, B_m$ are each determined by what happens on a different toss of a die, for the reason we discussed above in the case of two events, we take the events $B_1, B_2, \ldots, B_m$ to be independent.

Exercises

1. Prove: If A and B are mutually exclusive, and A and B are also independent, then either A or B has probability zero.

2. a. Prove: If A and B are independent, then so are A and $\overline{B}$.

   b. Prove: If A and B are independent, then so are $\overline{A}$ and $\overline{B}$.

   c. Give an example of events A, B, and C such that

$$P(A \cap B \cap C) = P(A)P(B)P(C),$$

   but

$$P(\overline{A} \cap \overline{B} \cap \overline{C}) \neq P(\overline{A})P(\overline{B})P(\overline{C}).$$

3. Prove:

   a. Suppose that

$$P(A \cap B \cap C) = P(A)P(B)P(C),$$
$$P(\overline{A} \cap B \cap C) = P(\overline{A})P(B)P(C),$$
$$P(A \cap \overline{B} \cap C) = P(A)P(\overline{B})P(C)$$

and

$$P(A \cap B \cap \overline{C}) = P(A)P(B)P(\overline{C}).$$

   Then, A, B, and C are independent.

   b. Suppose A, B, and C are independent. Then so are $\overline{A}$, B, and C.

   c. Suppose A, B, and C are independent. Then so are $\overline{A}$, $\overline{B}$, and C.

   d. Suppose A, B, and C are independent. Then so are $\overline{A}$, $\overline{B}$ and $\overline{C}$.

4. Prove: If

$$\frac{P(A)}{P(A \cap B)} + \frac{P(B)}{P(A \cap B)} = \frac{1}{P(A)} + \frac{1}{P(B)},$$

   then A and B are independent.

5. A card is drawn from a deck and then replaced; then a second card is drawn. Let:

> A be "The first card is a spade."
> B be "The second card is a spade."
> C be "Both cards are the same color."

Are A and B independent? How about B and C? A, B, and C?

6. In this problem, we say a die falls "high" if it falls "four," "five," or "six"; otherwise, we say it falls "low." Two dice are thrown. Let:

> A be "The first die falls high."
> B be "The second die falls high."
> C be "One die falls high and one low."
> D be "The total is four."
> E be "The total is five."
> F be "The total is seven."

Which of the following statements are true?

a. A and F are independent.

b. A and D are independent.

c. A and E are independent.

d. $P(A \cap B \cap C) = P(A)P(B)P(C)$.

e. A and C are independent.

f. C and E are independent.

g. $P(A \cap C \cap E) = P(A)P(C)P(E)$.

h. A, C, and E are independent.

7. For each of $n = 4, 5, 6, 7$, what is the probability that the World Series lasts just n games if the games are independent and, in each game, the probability that the National League team wins is: a) .5? b) .6?

8. Four passengers enter an elevator. There are four floors at which they can get off. The floors at which the passengers get off are independent. Each passenger is as likely to get off at any one floor as at any other floor. What is the probability that:

a. All passengers get off at the same floor?

b. All get off at different floors?

    c. Three get off at one floor and one at another?

    d. Two get off at one floor and two at another?

9. A snack bar sells three flavors of ice cream cones. Customers choose the flavor they want independently. The probabilities that a random customer orders vanilla, chocolate, and strawberry are 1/2, 1/3, and 1/6, respectively. What is the probability that three customers order three different flavors?

10. A coin is tossed repeatedly until tails has occurred ten times. a) What is the probability that, at that point, heads has not occurred twice in a row? b) What is the probability that, at that point, tails has not occurred twice in a row?

11. Four dice are thrown.

    a. What is the probability that none of them fall higher than "three"?

    b. What is the probability that none of them fall higher than "four"?

    c. What is the probability that "four" is the highest number thrown?

12. Five dice are thrown.

    a. What is the probability that none of them fall "one"?

    b. What is the probability that neither "one" nor "two" is thrown?

    c. What is the probability that "two" is the lowest number thrown?

**13. Four dice are thrown. What is the probability that "five" is the highest number thrown and "three" is the lowest?

**14. Five dice are thrown. What is the probability that "two" is the lowest number thrown and "six" is the highest?

3.2 Bernoulli Trials

A certain situation will arise many times in the future. To make it easy to announce that we are making the same basic assumptions that we regularly make, we introduce the term, "Bernoulli trials."

[The Bernoulli family included at least a dozen mathematicians; here we refer to Jakob Bernoulli (1654–1705). We shall say more about Jakob and the other Bernoullis in Chapter Five.] Frequently, we are concerned with an experiment in which something is done more than once. We speak then of repeated trials. Examples are obvious: We could toss a pair of dice 27 times. Or, we could throw one coin until it falls heads for the fifth time. In fact, every problem we have considered could be made into an example of repeated trials just by deciding to repeat many times whatever was done. We still have just one sample space; the points of this one sample space each describe how each of the trials comes out. We call the trials *Bernoulli trials* when three conditions are satisfied:

1. The trials are "independent"; the way any one trial turns out does not affect the chances of any outcome on any other trial.

2. We designate a certain outcome of each trial as "success"; when this outcome does not occur, we say the trial resulted in "failure."

3. The probability of success on any one trial is the same as the probability of success on any other trial.

For example, in repeated throws of a pair of dice, throwing seven or higher could be called a success. Or, in tossing a coin, a head could be a success. By convention, we use the letter p for the probability, mentioned in 3 above, of success on any one particular trial. We always let $q = 1 - p$. To avoid triviality, we always assume that $0 < p < 1$. (The definition of Bernoulli trials just given is obviously rather informal. It is possible to be more abstract, but only at the risk of being mysterious and hiding the intuitive ideas. The important thing is that we have independent events that all have the same probability.)

There are two obvious ways which we can set up an experiment involving Bernoulli trials. We can decide before starting the experiment how many trials we intend to make; say we decide on n trials. Then we can ask, what is the probability that we shall get just k successes in these n trials? Or we can reverse the procedure. We can decide in advance how many successes we want and plan to keep trying until we get that number of successes; say we decide on r successes. Then we can ask, what is the probability that it will take just k trials to get these r successes? Let us compute these probabilities.

We now assume that the number n of trials is fixed in advance. We seek the probability that k of these n trials result in success. First, what is the probability that the first k trials each result in success? Clearly p^k since the trials are independent and we have the probability p of success for each particular trial. Assuming that the first k trials do all result in success, we must have a failure on every one of the remaining $n - k$ trials if the total number of successes is to be k. The probability that all the last $n - k$ trials result in failure is q^{n-k}. We use independence to combine these conclusions; $p^k q^{n-k}$ is the probability that the first k trials all give success and the last $n - k$ trials give failure. If we select in advance any other k trials, the probability of success on all the selected trials and failure on all the other trials is also $p^k q^{n-k}$. We are trying to find the probability of getting exactly k successes, without regard to which trials they occur on. To get that probability, we add up the probabilities of k successes on specified trials for each choice of k trials. There are

$$\binom{n}{k}$$

ways to choose k trials out of n, and $p^k q^{n-k}$ is the probability of success on the chosen trials only for each choice. Thus

$$\binom{n}{k} p^k q^{n-k}$$

is the probability we seek. It is clear from the context that the n we have been using is a positive integer. Our reasoning applies when k is any integer with $0 < k < n$. Since we made the obvious definition

$$\binom{n}{0} - 1,$$

the conclusion obviously holds for $k = 0$ and $k = n$. Thus, for any integers n and k with $0 \leq k \leq n$, the probability of k successes in n Bernoulli trials is

$$\binom{n}{k} p^k q^{n-k}.$$

Now we consider the other problem. Suppose we plan to continue until we get r successes. What happens if we keep going and going and never get r successes? In the first place, we shall see later that this cannot happen; sooner or later we will have the r successes. In the second place, the question is irrelevant anyhow. We want to find the probability that it takes just k trials to get the r successes. After the kth trial, we shall know whether or not that event happened. To say that it takes just k trials to get r successes is to say that the rth success occurs on the kth trial. Rewording our requirements once more, we see we must have a success, the rth success, on the kth trial, and we must have had $r - 1$ successes on the earlier trials. We finally get to the point: It takes k trials to get r successes exactly when there are $r - 1$ successes on the first $k - 1$ trials and then a success on the kth trial. As we saw in the last paragraph, the probability of $r - 1$ successes in $k - 1$ trials is

$$\binom{k-1}{r-1}p^{r-1}q^{k-1}.$$

The probability of success on the kth trial is p. The events whose probability we just announced are independent, since one event refers to the first $k-1$ trials and the other to the kth trial. Thus we may multiply these probabilities to get our final conclusion. However we first must make explicit the values of r and k for which our reasoning holds; from the context, r and k must be positive integers with $r \leq k$. For such r and k, the probability that it takes k trials to get r successes is

$$\binom{k-1}{r-1}p^r q^{k-r}.$$

Exercises

15. A coin is tossed repeatedly until it falls heads for the fourth time. What is the probability that the fourth head occurs on the seventh toss?

16. If 101 coins are tossed, what is the probability that at least 51 fall heads?

17. A die is thrown four times.

   a. What is the probability it falls "six" just once?

   b. What is the probability it falls "six" at least once?

   c. What is the probability it falls "six" for the first time on the last throw?

18. A marksman scores a bull's eye on 90% of his shots.

   a. What is the probability that he gets at least eight bull's eyes if he shoots ten times?

   b. If he shoots until he gets eight bull's eyes, what is the probability that he needs at most ten shots?

19. If nine coins are tossed, what is the probability that the number of heads is even? How about 99 coins? 100 coins?

20. Consider the following two systems of Bernoulli trials:

   1. A coin is tossed; heads is a success.

   2. A die is thrown; "six" is a success.

   a. For each of 1 and 2, find the ratio $P(A)/P(B)$, where:

         A is "The third success occurs on the fifth trial."

         B is "Three of the first five trials result in success."

   b. Generalize, replacing three by i and five by j.

21. Samuel Pepys described dice labeled with the letters A, B, C, D, E, F instead of the numbers 1, 2, 3, 4, 5, 6. Then he posed the following question to Isaac Newton: "Peter a criminal convict being doomed to dye, Paul his friend prevails for his having the benefitt of one throw only for his life, upon dice soe prepared; with the choice of any one of these three chances for it, viz.,

One F at least upon six such dice.
Two F's at least upon twelve such dice.
or
Three F's at least upon eighteen such dice.

Question: - Which one of these chances should Peter in this case choose?"

(Suggestion: Use a calculator, even though Newton didn't have one.)

**22. Two men, Ben and Frank, want to settle who will pay a dinner check, which must be placed on a single credit card. To prolong the excitement, they do not want to decide on the basis of a single toss of a coin. Ben suggests that they each toss a coin 20 times and the one who gets the most heads wins. Frank points out that there could be a tie. Ben then proposes the following: Frank wins in case they toss the same number of heads, but Ben will get to toss 21 times to Frank's 20. Is this fair?

3.3   The Most Likely Number of Successes

This section may be postponed indefinitely.
It is not needed in the rest of the book.

An obvious question is, "In a given number of Bernoulli trials, with a given probability of success on each trial, what is the most likely number of successes?" In other words, if, for some fixed n, we set

$$a_k = \binom{n}{k} p^k q^{n-k},$$

for which k is a_k largest? Towards finding an answer, we compare a_k to a_{k-1}. Whether a_k is larger or smaller than a_{k-1} corresponds to whether a_k/a_{k-1} is larger or smaller than 1. We have

$$
\frac{a_k}{a_{k-1}} = \frac{\dfrac{n!}{k!(n-k)}p^k q^{n-k}}{\dfrac{n!}{(k-1)!(n-k+1)!}p^{k-1}q^{n-k+1}}
$$

$$
= \frac{(k-1)!}{k!} \frac{(n-k+1)!}{(n-k)!} \frac{p}{q}
$$

$$
= \frac{n-k+1}{k} \frac{p}{q}
$$

$$
= \frac{np - kp + p}{kq}.
$$

This fraction is larger than 1 exactly when its numerator $np - kp + p$ is larger than its denominator kq, in other words, when $(np - kp + p) - (kq)$ is larger than 0. We have

$$(np - kp + p) - (kq) = np - k(p + q) + p = np - k + p = p(n + 1) - k.$$

Thus whether a_k/a_{k-1} is larger than 1 is determined by whether $p(n + 1)$ is larger than k. To summarize,

$$\text{if } k < p(n + 1), \text{ then } a_k > a_{k+1};$$

$$\text{if } k > p(n + 1), \text{ then } a_k < a_{k+1}.$$

In the exceptional case $k = p(n + 1)$, $a_k = a_{k+1}$. First consider the case where $p(n+1)$ is not an integer; then $k = p(n+1)$ is impossible. Looking at $a_0, a_1, \ldots, a_n$ in turn, we find that each a_k is larger than the one before as long as $k < p(n + 1)$; after that point, each a_k is smaller than the one before. The k for which a_k is largest is thus the largest integer less than $p(n + 1)$. With the usual notation $[x]$ for the largest integer not exceeding x, we have shown: The most likely number of successes is $[p(n+1)]$, provided $p(n+1)$ is not an integer. Now consider the other case where $p(n + 1)$ is an integer. Then, for $k = p(n + 1)$, $a_k = a_{k-1}$. Reasoning as before, we see that $p(n + 1)$ and $p(n + 1) - 1$ are tied in the race to be the most likely number of successes. There is just as much chance of $p(n + 1)$ successes as $p(n + 1) - 1$ successes, and more chance of these numbers of successes than any other number of successes.

We give an example. Suppose $p = 4/5$ and $n = 8$. We list in a table the probability of k successes in eight Bernoulli trials with probability

k	Probability of k successes
0	.0000026
1	.0000819
2	.0011469
3	.0091750
4	.0458752
5	.1468006
6	.2936013
7	.3355443⇐⇐⇐⇐⇐
8	.1677722

4/5 of success on each trial. Of course, the numbers in the table are approximations accurate to the number of decimal places shown. The most likely number of successes is clearly seven, and we have $[p(n+1)] = [7.2] = 7$.

Exercises

23. n coins are tossed. For each of $n = 100$, 101, 102, and 103, what is the most likely number of coins to fall heads?

24. n dice are thrown. For each of $n = 14$, 15, 16, 17, 18, and 19, what is the most likely number of dice to fall "six"?

25. Twenty-two red dice and ten blue dice are thrown. What is the most likely number of "sixes" to be shown on:

    a. the red dice?

    b. the blue dice?

    c. all the dice?

26. n pairs of dice are thrown. For each of $n = 2$, 4, 6, and 8, what is the most likely number of pairs to total six?

27. Urn A contains two red balls and eight blue balls. Urn B contains two red balls and ten green balls. Six balls are drawn from urn A and four are drawn from urn B; in each case, each ball is replaced before the next one is drawn. What is the most likely number of blue balls to be drawn? What is the most likely number of green balls to be drawn?

28. Prove: In a system of Bernoulli trials, the first success is more likely to occur on the first trial than on any other trial.

29. Prove: Consider a system of Bernoulli trials and a fixed positive integer s. Let

$$x = \left[\frac{s-q}{p}\right].$$

With the exception noted below, the sth success is more likely to occur on the xth trial than on any other trial. However, if

$(s - q)/p$ is an integer ≥ 2, the sth success is as likely to occur on the $(x - 1)$st trial as on the xth trial.

3.4 Conditional Probability

Our judgement as to how likely a certain event is may be influenced by the knowledge that some other event definitely takes place. Before making a formal definition, we discuss the intuitive ideas behind the definition. First we raise a question. Suppose we know that an event B has happened. How likely is it that A happens under these circumstances? In other words, in many repetitions of our experiment, what fraction of these occasions on which B happens does A also happen? To make our ideas more vivid, we use an example.

In fact, we'll work a problem before developing the theory behind it. A certain company classifies absenteeism as high when more than 12% of its employees are absent. Absenteeism is high on 1/18 of all Tuesdays, Wednesdays, and Thursdays. However, absenteeism is high on 1/6 of all Mondays and Fridays. A distinguished visitor is going to visit the company on a day yet to be determined. This day is as likely to be any one weekday as any other. What is the probability that absenteeism is high on the day the visitor comes? We first compute the probability that the visitor comes on a Monday or Friday and absenteeism is high on that day. Two-fifths of all weekdays are Mondays or Fridays. Of this 2/5 of all weekdays, 1/6 are days with high absenteeism. Thus $(1/6)(2/5) = 1/15$ of all weekdays have high absenteeism and are Mondays or Fridays. Likewise, $(1/18)(3/5) = 1/30$ of all weekdays have high absenteeism and are Tuesdays, Wednesdays, or Thursdays. Thus the probability of high absenteeism on a randomly chosen weekday is $1/15 + 1/30 = 1/10$. That answers the question above.

Strictly speaking, the numbers 1/6 and 1/18 in the problem just worked are not probabilities. The experiment was to choose a day, and nothing in particular happened on 1/6 of *all* days. One-sixth is the "probability" of one event, high absenteeism, on the assumption that another event, Monday or Friday, occurred. In other words,

$$\frac{1}{6} = \frac{1/15}{2/5}$$

is the ratio of the number of days that both have high absenteeism and are Monday or Friday to the number of days that are Monday or Friday.

Let us now generalize. Let A and B be events. We define

$$P(A \mid B) = \frac{P(A \cap B)}{P(B)}$$

This equation makes sense only when $P(B) \neq 0$, since we cannot divide by zero. Thus, $P(A \mid B)$ is defined only when $P(B) \neq 0$. We read $P(A \mid B)$ as "the probability of A given B." We call $P(A \mid B)$ a *conditional probability*. The intuitive idea behind the definition is that, in many repetitions of our experiment, $P(A \mid B)$ is the fraction of those occasions on which B happens that A also happens.

It might seem appropriate to discuss the history of "conditional probability" here. But, in fact, there really is no such history. In a certain sense, all probabilities are conditional. We ask about, for example, the probability that three coins will fall heads, *under the condition* that five coins are tossed. Alternatively, we could ask about the probability that three coins fall heads, under both of the conditions that five coins are tossed and that at least two fall heads. In more philosophical discussions, all probabilities are shown in the form $P(A \mid B)$ for this reason. Of course, from the point of view of abstract mathematics, we can clearly see that $P(A)$ depends on one subset of the sample space while $P(A \mid B)$ depends on two. But this twentieth-century abstraction is irrelevant to the history of the beginnings of probability theory. The discussion of Fermat and Pascal involved conditional probability from the first; they simply did not distinguish conditional probability from probability.

Suppose A and B are independent and $P(B) \neq 0$. By definition, we have

$$P(A \mid B) = \frac{P(A \cap B)}{P(B)}.$$

Since A and B are independent,

$$\frac{P(A \cap B)}{P(B)} = \frac{P(A)P(B)}{P(B)}$$
$$= P(A).$$

Thus for independent A and B, $P(A \mid B) = P(A)$. Conversely, if $P(A \mid B) = P(A)$, it follows that A and B are independent. Note that if neither $P(A)$ nor $P(B)$ is zero, then $P(A \mid B) = P(A)$ holds exactly when $P(B \mid A) = P(B)$.

From the definition of $P(A \mid B)$, we have

$$P(A \cap B) = P(B)P(A \mid B).$$

If any two of the three numbers $P(A \cap B)$, $P(A \mid B)$, and $P(B)$ are known, the equation may be used to find the third of the numbers. In particular, it may be used to find $P(A \cap B)$ even when A and B are not independent. Before illustrating this use with examples, we state the last equation in words. We can probably do no better than to quote, from *The Doctrine of Chances* by Abraham DeMoivre, the first explicit statement of the idea of the equation. (We shall say more about DeMoivre and his book in Chapter Six.) As DeMoivre says, "... the Probability of two Events dependent, is the product of the Probability of the happening of one of them, by the Probability which the other will have of happening, when the first shall have been consider'd as having happen'd; and the same rule will extend to the happening of as many Events as may be assign'd." The meaning of the final comment about extending the rule will be made clear in the examples to which we now proceed.

1. Suppose two cards are drawn from the deck, without replacement. What is the probability that both are aces? Of course, this problem may be solved by the methods of the last chapter. But we now have an alternative method that simplifies the computation. We may suppose the cards to be drawn one at a time. Let A be the event, "The first card is an ace." Let B be "The second card is an ace." We seek $P(A \cap B)$. We have $P(A \cap B) = P(B \mid A)P(A)$. Note $P(A) = 4/52$. After the first card is drawn, the second card is chosen from the remaining 51 cards. If the first card is an ace,

there are 3 aces left among the 51 cards. Thus, $P(B \mid A) = 3/51$. Hence $P(A \cap B) = (4/52)(3/51) = 1/221$.

2. Suppose three cards are drawn from the deck, without replacement. What is the probability all three are aces? Let C be "The first two cards are both aces." Let D be "The third card is an ace." We seek $P(C \cap D) = P(D \mid C)P(C)$. By the last example, $P(C) = (4/52)(3/51)$. If the first two cards are aces, the third card is chosen from 50 cards, of which 2 are aces. Thus, $P(D \mid C) = 2/50$. Hence $P(C \cap D) = (4/52)(3/51)(2/50) = 1/5525$.

3. What is the probability that all five cards in a poker hand are spades? Since the problem is like the ones we just did about aces, we shall be brief. The probability the first card is a spade is

$$\frac{13}{52}.$$

The probability that the first two cards are both spades is

$$\frac{13}{52} \cdot \frac{12}{51}.$$

The probability the first three cards are all spades is

$$\frac{13}{52} \cdot \frac{12}{51} \cdot \frac{11}{50}.$$

The probability the first four cards are all spades is

$$\frac{13}{52} \cdot \frac{12}{51} \cdot \frac{11}{50} \cdot \frac{10}{49}.$$

Thus the answer to the original question is

$$\frac{13}{52} \cdot \frac{12}{51} \cdot \frac{11}{50} \cdot \frac{10}{49} \cdot \frac{9}{48} = \frac{33}{66640}.$$

Note that this answer could have been written down directly if we had omitted the explanation.

4. Given that a poker hand (five cards) contains both black aces, what is the probability that it contains at least three aces? We work this problem by several different methods to illustrate both the methods and the flexibility of the concept of conditional probability. The first method by which we solve the problem begins by defining certain events:

A is "The hand contains both black aces."

B is "The hand contains at least three aces."

Thus we seek $P(B \mid A)$. We have

$$P(B \mid A) = \frac{P(A \cap B)}{P(A)}.$$

We find $P(A)$ first. How many poker hands contain both black aces? As many as there are ways to choose the other three cards the hand contains in addition to the black aces. Thus there are

$$\binom{50}{3} = \frac{50 \cdot 49 \cdot 48}{3 \cdot 2} = 19{,}600$$

hands with both black aces; hence

$$P(A) = \frac{19{,}600}{\binom{52}{3}}.$$

Now consider $P(A \cap B)$. We need hands with both black aces and at least one additional ace. That includes the 48 hands with all four aces. To get the hands with both black aces and one red ace, we must pick one red ace out of two red aces and two non-aces out of 48 non-aces. Thus there are

$$48 + 2\binom{48}{2} = 48 + 48 \cdot 47 = 2304$$

poker hands containing both black aces and at least three aces in all. Thus

$$P(A \cap B) = \frac{2304}{\binom{52}{5}}.$$

When we divide $P(A \cap B)$ by $P(A)$, the denominators of $\binom{52}{5}$ cancel out and we get

$$\frac{2304}{19600} = \frac{144}{1225}$$

as our answer.

Our second method is basically the same as the first. The difference is that we anticipate the fact that the total number of poker hands, $\binom{52}{5}$ will cancel out. We simply compute the number, 19,600, of hands that contain both black aces and determine that 2304 of these hands contain at least three aces. Thus 2304/19,600 gives us our answer.

The third method is an obvious modification of the second. In the last paragraph, we particularly studied the three cards in the poker hand besides the black aces. We now restrict our attention entirely to these three cards. These three cards can be any of the 50 cards left when the black aces are excluded. The three cards are as likely to be any three of the 50 as any other three of the 50. Thus we are trying to find the probability that when three cards are chosen from a certain 50, at least one ace is chosen. The obvious way to find that probability is to begin by finding the probability that no aces are chosen. Reasoning in this way, we find the answer to our original question to be

$$1 - \frac{\binom{48}{3}}{\binom{50}{3}}.$$

Again we change our approach slightly. As in the last method, we are concerned with choosing three of the 50 cards left when both black aces are excluded. We may suppose the cards are to be selected one at a time. As before, we first compute the probability that none of the chosen cards is an ace. The probability the first card chosen is not an ace is 48/50, since only two of the 50 cards are aces. Assuming that the first card is not an ace, the probability the second card is not an ace is 47/49. Likewise we have 46/48 for the third. Thus our answer is

$$1 - \frac{48}{50} \cdot \frac{47}{49} \cdot \frac{46}{48}.$$

Which of the four methods is best? The one we think of first. None is overly complicated. Thus, if we have one method in hand, looking for others merely gives us additional work.

Suppose we are trying to estimate how likely a certain event E is. We may say to ourselves, "If such and such happens, then the chance of E happening is thus and so. On the other hand, under the following different circumstances, the chance of E is this or that." If we choose to distinguish more than two alternative conditions, we would continue in the same vein. The basic procedure is to identify various mutually exclusive sets of circumstances that together cover all possibilities. Then we estimate the likelihood of each set of circumstances and the chances for E under each of them. The next theorem tells us how to put all these estimates together.

Before stating the theorem, we introduce some terminology that we shall use here and elsewhere. We call $A_1, A_2, \ldots, A_n$ a *partition* of Ω if each A_i is an event, and, whenever the experiment is done, just one of the events A_i must happen. Thus, in the first place, $A_1, \ldots, A_n$ must be mutually exclusive; i.e., $A_i \cap A_j = \emptyset$ whenever $i \neq j$. And, in the second place, $A_1 \cup \cdots \cup A_n = \Omega$. The following theorem is almost obvious, but we state it here because it is so useful.

Theorem *Let $A_1, \ldots, A_n$ be a partition of Ω. Suppose none of the A_i have probability zero. Let B be any event. Then*

$$P(B) = P(B \mid A_1)P(A_1) + P(B \mid A_2)P(A_2) + \cdots + P(B \mid A_n)P(A_n).$$

Proof For each i, we have $P(A_i \cap B) = P(B \mid A_i)P(A_i)$. Also, since $A_1 \cap B, \ldots, A_n \cap B$ are mutually exclusive and $B = (A_1 \cap B) \cup \cdots \cup (A_n \cap B)$,

$$P(B) = P(A_1 \cap B) + \cdots + P(A_n \cap B).$$

$\square$

We illustrate with an example. On a television show, a contestant is required to throw a die. She then tosses as many coins as there were spots shown on the die. If she throws exactly four heads, she wins \$10,000. What is the probability she wins? Two possible choices for a partition suggest themselves. We could use $A_1, A_2, \ldots, A_6$, corresponding to the six ways the die might fall. Instead we define events B, A_1, A_2, A_3, A_4 by

B is "The contestant gets the \$10,000."

A_1 is "The die falls 'one', 'two' or 'three'."

A_2 is "The die falls 'four'."

A_3 is "The die falls 'five'."

A_4 is "The die falls 'six'."

$P(B \mid A_1)$ is clearly 0, since one cannot get four heads by tossing three or fewer coins.

$$P(B \mid A_2) = \frac{1}{16},$$

the probability of four heads when four coins are tossed. Likewise we compute

$$P(B \mid A_3) = \binom{5}{4}\frac{1}{32} = \frac{5}{32},$$

$$P(B \mid A_4) = \binom{6}{4}\frac{1}{64} = \frac{15}{64}.$$

Obviously, $P(A_1) = 1/2$ and $P(A_2) = P(A_3) = P(A_4) = 1/6$. Thus we have

$$P(B) = P(B \mid A_1)P(A_1) + P(B \mid A_2)P(A_2) + P(B \mid A_3)P(A_3)$$
$$+ P(B \mid A_4)P(A_4)$$
$$= 0 \cdot \frac{1}{2} + \frac{1}{16}\cdot\frac{1}{6} + \frac{5}{32}\cdot\frac{1}{6} + \frac{15}{64}\cdot\frac{1}{6}$$
$$= \frac{29}{384}.$$

The following theorem goes just one step further than the last theorem. It is a matter of taste whether one should memorize and apply the theorem, or avoid possible faulty recollection by simply thinking things out. The theorem does have a name, "Bayes's Formula"; it was named after Thomas Bayes, whom we will discuss shortly. It does solve a natural problem: Suppose we know how likely an event B is under each of various sets of circumstances. Assume that we have an opinion as to how likely it is for each of these sets of circumstances to occur. How do we modify our opinion if we learn that B does in fact occur?

Theorem *Let $A_1, A_2, \ldots, A_n$ be a partition of Ω. Suppose none of the A_i have probability zero. Let B be an event with $P(B) \neq 0$. Then, for* *each i,*

$$P(A_i \mid B) = \frac{P(B \mid A_i)P(A_i)}{P(B \mid A_1)P(A_1) + P(B \mid A_2)P(A_2) + \cdots + P(B \mid A_n)P(A_n)}.$$

Proof From the definition of conditional probability we have

$$P(A_i \mid B) = \frac{P(A_i \cap B)}{P(B)},$$
$$P(A_i \cap B) = P(B \mid A_i)P(A_i).$$

We use the last theorem to evaluate $P(B)$. □

We apply Bayes's Formula to the situation of the television contestant in the last paragraph. We consider the following problem: If the contestant wins the \$10,000, what is the probability that the die fell "four"? With notation as before, we seek $P(A_2 \mid B)$. The theorem states that

$$P(A_2 \mid B)$$
$$= \frac{P(B \mid A_2)P(A_2)}{P(B \mid A_1)P(A_1) + P(B \mid A_2)P(A_2) + P(B \mid A_3)P(A_3) + P(B \mid A_4)P(A_4)}$$

Thus we have

$$P(A_2 \mid B) = \frac{\dfrac{1}{16} \cdot \dfrac{1}{6}}{0 \cdot \dfrac{1}{2} + \dfrac{1}{16} \cdot \dfrac{1}{6} + \dfrac{5}{32} \cdot \dfrac{1}{6} + \dfrac{15}{64} \cdot \dfrac{1}{6}}$$
$$= \frac{4}{29}.$$

We give another application of Bayes's Formula. Even when a gizmo-making machine is in good working order it occasionally malfunctions; each gizmo made has independently a 1% chance to be defective. A certain gizmo-making machine has just had a new part installed. The first gizmo it makes is defective. It is known that one in every 50 new parts installed is a dud. When a dud is used in a gizmo-making machine, only 10% of its output is usable. Should the new part be replaced?

All we can do is compute the probability that the part is a dud. Let us introduce some events:

D is "The new part is a dud."

A is "The new part is all right."

F is "The first gizmo made is defective."

We are told that $P(D) = .02$; hence $P(A) = .98$. We also have $P(F \mid D) = .9$ and $P(F \mid A) = .01$. We seek $P(D \mid F)$. Bayes's Formula states that

$$P(D \mid F) = \frac{P(F \mid D)P(D)}{P(F \mid D)P(D) + P(F \mid A)P(A)}.$$

All we need to do is substitute. We have

$$P(D \mid F) = \frac{(.9)(.02)}{(.9)(.02) + (.01)(.98)}$$
$$= \frac{.018}{.018 + .0098}$$
$$= .6475$$

to four places; that is, roughly two-thirds. While the part is more likely than not to be a dud, there is about one chance in three that it is all right.

We conclude the chapter with some comments about history. First we discuss Bayes and what Bayes actually did. Bayes wrote a single, but very important, paper involving probability. In the paper, he considered a certain problem in statistics; in doing so, he founded the subject of statistics. The problem is easy for us to describe here: Suppose we do an experiment involving Bernoulli trials. Assume that we do not know the probability p of success on any one trial. How do we best estimate p from the number of successes that actually occur? Of course, we cannot find p; the observed number of successes could have been obtained for any p. But it may well be that that number would have been very unlikely to occur for certain values of p. Bayes describes a method that he claims may be used to determine the probability that p lies in any given interval. The method is essentially equivalent to the application of a special case of the theorem we have called Bayes's Formula. Now Bayes's problem involves continuous, rather than discrete, probability; the p involved can vary continuously from zero to one with every value in between being a possibility. Towards continuing to try to understand what Bayes did, let us replace his problem with a discrete, if artificial, problem. Suppose we make a known number n of Bernoulli trials, knowing only that the probability p of success on a single trial is one of the numbers $.1, .2, .3, \ldots, .9$. Assume we observe that s successes occur. For each of the listed possibilities for p, what is the probability that p had that value? Trying to use Bayes's formula, as we stated it, we see that we need to know $P(A_1), \ldots, P(A_9)$, the probabilities we would have assigned to each possible value for p before doing the experiment. Bayes's paper proceeds as though we could simply cancel out all these $P(A_i)$ on the ground that they are all the same, namely unknown. (Later work in this area treats the question of assigning "prior" probabilities in more detail.) Thus Bayes's formula, as he stated it, is the continuous analog of the special case of our Bayes's Theorem where $P(A_1) = \cdots = P(A_n)$.

Thomas Bayes

The Reverend Thomas Bayes (English, 1702–1761)

Thomas Bayes's father, Joshua, was one of the first six men to be publicly ordained Nonconformist ministers. That is to say, when it became legal to be a minister outside of the established Anglican Church, Joshua Bayes was one of the first such ministers. He was a respected theologian and also a fellow of the Royal Society. Thomas followed in his father's footsteps, becoming minister of the Presbyterian Chapel in Tunbridge Wells. Thomas Bayes published two works in his lifetime. The first was about theology. The second, published anonymously, was an attempt to refute Bishop Berkeley's objections to calculus as illogical (George Berkeley, 1685–1753). Bayes did the best that could be done at the time; it was not until the late nineteenth century that the reasoning behind calculus was brought up to modern standards. Bayes was elected a fellow of the Royal Society in 1742. He retired a rich man in 1750. His most important work, the "Essay Towards Solving a Problem in the Doctrine of Chances," was not published until 1763, but was judged important enough to be worth republishing for its ideas as recently as 1958.

We continue our discussion of the history of probability theory by turning to the work of Laplace. The importance of Laplace as a scientist can best be judged by his omission from the *Dictionary of Scientific Biography* (published by Scribner's), as it first appeared. Someone was, of course, selected to write up a biography of Laplace, which was to include a summary of Laplace's work. When the time to publish arrived, no biography had been completed. Thus it was necessary to find a new biographer and leave Laplace to appear in the scheduled supplementary volume. But when the time came to print this supplement, no biography had yet been completed. After "much delay" had been caused to the supplement, the editor despaired of any one person's being able to summarize Laplace's work; accordingly the task was divided into five parts, each for a different person. The best we can do here is to list some of the topics on which Laplace worked: shape of the earth, comets, tides, chemical physics of heat, moons of Jupiter, our moon, the velocity of sound.

Laplace published the first edition of his short popular book, *Essai philosophique sur les probabilités*, in 1814. He also published four revisions, the last in 1825. In the book, he presents a summary of the rules for computing probabilities that we have developed in our book so far. He includes Bayes's Formula, essentially as stated above. Laplace's book is very much concerned with applications of probability to human affairs. In general, Laplace was most interested in applying probability theory to his scientific work. His contributions to probability theory were primarily technical methods of computing probabilities and are beyond the scope of this book.

Laplace

Pierre Simon, Marquis de Laplace (French, 1749–1827)

Laplace was born in Normandy; reports about the circumstances of his family are conflicting. Laplace attended a school where half the students had their expenses paid for by the king; most students went on to the church or the army. Laplace demonstrated aptitude in all areas and particularly distinguished himself in debating the subtle points of theology. Nevertheless, his primary interest was mathematics. Upon completing his studies at the school, he immediately secured a position teaching mathematics there. At the age of 18 or 19, Laplace decided to visit Paris; he took with him letters of introduction. Despite the letters, he was unable to secure a meeting with Jean d'Alembert (1717–1783), the leading French mathematician of the day. Then Laplace himself wrote a letter to d'Alembert discussing mechanics. D'Alembert was so struck by this letter that, the day he received it, he arranged to meet Laplace. D'Alembert was able to secure for Laplace the choice between professorship in Berlin or at the Ecole Militaire in Paris. Laplace preferred the latter. He became a member of the Academy of Sciences in 1795. Laplace also became a confidant of Napoleon. When Napoleon came to power, he gave the position of Minister of the Interior to Laplace. Napoleon wrote later, "Laplace was a worse than poor administrator. He considered nothing beneath his notice. He searched for subtleties everywhere and everywhere looked only for the infinitely small." Laplace was replaced after six weeks. However he was made a senator and became President of the Senate. In 1806, he was made a count. Despite this, in 1814 Laplace was one of the first to vote to oust Napoleon. When the Senate of the empire was replaced by the Chamber of Peers of the restored monarchy in 1817, Laplace was reclassified from count to the higher position of marquis. He continued his scientific work almost until his death in 1827. In 1842, the Chamber of Deputies passed a law providing for the printing of a new edition of Laplace's work at the expense of the state.

Exercises

30. Suppose A, B, and C are events none of whose probabilities are zero. Prove:

    a. $P(\overline{A} \cup \overline{B}) = 1 - P(B)P(A \mid B)$.

    b. $P(A \cap B \mid B \cup C) = P(A \cap B \mid B)P(B \mid B \cup C)$.

    c. $P(B \cap C \mid A) = P(C \mid A)P(B \mid A \cap C)$ if $P(A \cap C) \neq 0$.

    d. $P(A \mid B)P(B \mid C)P(C \mid A) = P(B \mid A)P(C \mid B)P(A \mid C)$.

    e. $\dfrac{P(A \mid A \cup B)}{P(B \mid A \cup B)} = \dfrac{P(A)}{P(B)}$.

31. Prove: $\dfrac{P(\overline{B} \mid A)}{P(B)} + \dfrac{P(\overline{A})}{P(A)} = \dfrac{P(\overline{A} \mid B)}{P(A)} + \dfrac{P(\overline{B})}{P(B)}.$

32. Suppose A, B, and C are independent events and $P(C) \neq 0$. Prove:

    a. $P(A \cap B \mid C) = P(A \mid C)P(B \mid C)$.

    b. $P(A \cup B \mid C) = P(A \mid C) + P(B \mid C) - P(A \cap B \mid C)$.

33. Suppose A, B, and C are independent. Prove that $P(A \mid B \cap C) = P(A)$ provided $P(B \cap C) \neq 0$.

34. Assume that each child in a family is independently as likely to be a boy as a girl. Consider families with just two children.

    a. If at least one child is a boy, what is the probability both children are boys?

    b. If the older child is a girl, what is the probability both children are girls?

35. A die is tossed five times.

    a. Given that the die falls "six" at least once, what is the probability that it falls "six" at least twice?

    b. Given that the first throw was a "six," what is the probability that "six" was thrown at least twice?

36. We select from a deck of cards the four kings and the aces of spades, diamonds, and hearts. From these seven cards, three are chosen at random.

    a. What is the probability that at least one ace is chosen?

    b. What is the probability that the ace of spades is chosen?

    c. What is the probability that all three aces are chosen?

    d. If it is known that at least one ace was chosen, what is the probability that all three aces were chosen?

    e. If it is known that the ace of spades was chosen, what is the probability that all three aces were chosen?

37. If a poker hand (five cards) is known to contain at least three aces, what is the probability that it contains all four aces?

38. If a poker hand (five cards) is known to contain the aces of hearts, clubs, and diamonds, what is the probability that it contains the ace of spades?

39. Find the probability that a poker hand (five cards) contains both black aces given that it contains at least three aces.

40. In a bridge game, Players N and S each are dealt 13 cards. If Player N has exactly two aces in his hand, what is the probability that Player S has at least one ace?

41. Suppose that in a hog calling contest the more skillful participant always wins. Suppose that Alphonse, Beatrice, and Charlene are randomly chosen hog callers. If Alphonse beats Charlene in hog calling, what is the probability that Alphonse is a better hog caller than Beatrice?

42. A coin is tossed ten times. Given that heads occurred for the first time on the second toss, what is the probability that heads occurred for the fifth time on the last toss?

43. We select from a deck of cards the four kings and the four queens. From these eight cards, we draw one card at a time, without replacement, until all eight cards have been drawn. Find the probability that:

    a. All kings are drawn before the queen of spades.

    b. There is at least one queen that is drawn after all the kings.

    c. Each queen is drawn before each of the kings.

    d. Each queen is drawn before the king of the same suit.

    e. The last king to be drawn is the sixth card to be drawn.

44. A closet contains three pairs of black socks and two pairs of white socks, all new and all of the same size and style. Two socks are chosen at random. What is the probability that these two socks can be worn together without exciting comment?

45. Do the last problem with "socks" replaced by "shoes" throughout.

46. Each employee of a certain large hotel works five days a week, getting two consecutive days off. The days off of the various employees are scattered evenly through the calendar week.

    a. If Smith and Jones each work for the hotel, what is the probability that they have a day off in common?

    b. If Doe also works for the hotel, what is the probability that no two of these three persons have a day off in common?

  c. What is the probability that there is a day on which all three are off?

47. The four aces and the four kings are selected from a pack of cards. From these eight cards, four cards are chosen at random. What is the probability that one card of each suit is chosen?

48. Four dice are thrown. What is the probability of obtaining four different numbers?

49. A drawer contains ten pairs of gloves, no two pairs alike. Alice takes one glove; then Betty takes a glove; then Alice takes a second glove; finally Betty takes a second glove. What is the probability that:

  a. Alice has a pair of gloves?

  b. Betty has a pair of gloves?

  c. Neither has a pair?

50. A closet contains six different pairs of shoes. Five shoes are chosen at random. What is the probability that at least one pair of shoes is obtained?

51. Six men and six women sit at a round table so that no two men sit next to each other; otherwise, the seating is random. What is the probability that Ms. Smith and Mr. Jones, who are bitter enemies, sit next to each other?

52. At a camera factory, an inspector painstakingly checks 20 cameras and finds that three of them need adjustment before they can be shipped. Another employee carelessly mixes the cameras up so that no one knows which is which. Thus the inspector must recheck the cameras one at a time until he locates all the bad ones.

  a. What is the probability that no more than 17 cameras need to be rechecked?

  b. What is the probability that exactly 17 must be rechecked?

53. Urn N contains five red balls and ten blue balls. Urn D contains two red balls and six white balls. A nickel and a dime are tossed. If the nickel falls heads, a ball is chosen at random from Urn N; if the dime falls heads, from Urn D. (Thus if both coins fall

heads, two balls are drawn.) What is the probability that at least one red ball is drawn?

54. Urn H contains six red balls and four white balls. Urn T contains two red balls and three white balls. A coin is tossed. If it falls heads, a ball is chosen from Urn H. If the coin falls tails, a ball is chosen from Urn T.

   a. If the ball is chosen from Urn H, what is the probability that it is red?

   b. What is the probability that the chosen ball is red and is chosen from Urn H?

   c. What is the probability that the chosen ball is red?

   d. If a red ball is chosen, what is the probability that it came from Urn H?

55. Urn A contains two red balls and three white balls. Urn B contains five red balls and one blue ball. A ball is chosen at random from Urn A and placed into Urn B. Then a ball is chosen at random from Urn B.

   a. What is the probability that both balls are red?

   b. What is the probability that the second ball is red?

   c. Given that the second ball is red, what is the probability that the first ball was red?

   d. Given that the second ball is blue, what is the probability that the first ball was red?

56. Urn A contains six red balls and six blue balls. Urn B contains four red balls and 16 green balls. A die is thrown. If the die falls "six," a ball is chosen at random from Urn A. Otherwise, a ball is chosen from Urn B. If the chosen ball is red, what is the probability that the die fell "six"?

57. A medical test is not completely accurate. When people who have a certain disease are tested, 90% of them have a "positive" reaction. But 5% of people without the disease also have a "positive" reaction. In a certain city, 20% of the population have the disease. A person from this city is chosen at random and tested; if the reaction is "positive," what is the probability the person has the disease?

58. The WT Company turns out 10,000 widgets a day. As everyone knows, the most important part of a widget is the idge at its center. WT buys it idges from two other companies: Sertane Company can supply only 2,000 idges per day, but 99% of them work properly. The remaining 8,000 idges are purchased from Kwik Company; 10% of these are defective. Suppose a randomly chosen widget from WT has a defective idge. What is the probability that Kwik is responsible?

59. An inspector has the job of checking a screw-making machine at the start of each day. She finds that the machine needs repairs one day out of ten. When the machine does need repairs, all the screws it makes are defective. Even when the machine is working properly, 5% of the screws it makes are defective; these defective screws are randomly scattered through the day's output. Use a calculator to get approximate answers to the following questions. What is the probability that the machine is in good order if:

    a. The first screw the inspector tests is defective?

    b. The first two screws are both defective?

    c. The first three are all defective?

60. The word spelled HUMOR by a person from the United States is spelled HUMOUR by a person from Britain. At a party, two-thirds of the guests are from the United States and one-third from Britain. A randomly chosen guest writes the word, and a letter is chosen at random from the word as written.

    a. If this letter is a U, what is the probability that the guest is from Britain?

    b. If the letter is an H, what is the probability that the guest is from Britain?

61. Urn A contains two black balls, urn B contains two white balls, and urn C contains one ball of each color. An urn is chosen at random. A ball is drawn from the chosen urn and replaced; then again a ball is drawn from that urn and replaced. If both drawings result in black balls, what is the probability that a third drawing from the same urn will also yield a black ball?

4

CHAPTER

Random
Variables

Often the results of an experiment are quantitative, rather than qualitative; in other words, the outcome of the experiment is a number or numbers. Assume for the moment that the result of the experiment is a single number, say the amount of money we shall be paid as the result of a bet. Then we may want to take into account that, if we receive $50, that is exactly half of $100. For some purposes, we regard a probability of one-fifth of getting $100 as equivalent to a probability of two fifths of getting $50. We next introduce random variables into our study of probability to make it possible to take a quantitative point of view along those lines.

The language that, for historical reasons, is involved in discussing random variables can be confusing. However, if we face up to the problem, there is no real difficulty. The term "random variable" itself is the first obstacle; we shall see in a moment that, strictly speaking, it is not a variable and is not chosen at random. But there is good reason, besides long-standing custom, to use the phrase "random variable." The phrase denotes something that varies at random. For example, we could throw a lot of coins, of different values, and count how many fall heads. The number of coins falling heads would be determined by chance. If we repeat the experiment, we shall likely get a different number of heads. We shall say that the number of coins that fall heads is a random variable X. The value of the coins that fall heads would be another random variable Y. We should not let the need for a formal definition obscure the basic simplicity of the concept.

A function that assigns a number to each point in the sample space is called a *random variable*. (Calling a function a variable, or a variable a function, is not unusual, even if it is somewhat confusing.) One obvious way in which random variables arise is in the situation where a bet is made on the outcome of our experiment. Then the amount of money paid is a random variable X. In more detail, if the point u of the sample space is the one that occurs, the number $X(u)$ assigned by X to u is the amount actually paid. Some very important random variables came very close to being introduced in the last chapter. The number of successes in a predetermined number of Bernoulli trials is a random variable. So is the number of trials needed to get a predetermined number of successes.

We next introduce some notation that is almost self-explanatory. Suppose we are given a random variable X and a number t. When our experiment is done, we get a particular point u of the sample space, and X assigns a certain value $X(u)$ to this point u. Either $X(u)$ is t or it is not; we consider the probability that $X(u) = t$. We write $P(X = t)$ for this probability. The event A involved here is simply the set of those $u \in \Omega$ for which $X(u) = t$; $P(X = t)$ means, by definition, $P(A)$.

4.1   Expected Value and Variance

We shall soon consider certain sums that have one term for each point of Ω. If Ω has infinitely many points, we shall then have infinite series. Infinite series and sums of finitely many terms are alike in many ways. On the whole, it will make little difference whether the number of points in Ω is finite. However, there are some important differences between infinite series and finite sums; most notably, infinite series sometimes diverge. So that we may concentrate on the important ideas without being distracted by continually discussing convergence, we first consider the case where Ω contains only finitely many points. Then later we can modify our work to cover the infinite case. The changes will be minor; as long as the series involved do converge, the same conclusions remain valid. Therefore, we now make the assumption, to remain in effect until further notice, that Ω contains only finitely many points.

Suppose we consider a random variable X on our sample space Ω. If we do the experiment just once, we get one point u of Ω, and X assigns a certain value $X(u)$ to this point. The value $X(u)$ cannot in general be predicted in advance. But what happens if we do the experiment many times and average the results?

We first discuss this averaging in an example. Suppose we make a bet as follows: We shall pay $20 in advance. Then we shall throw a die. The amount we are to be paid depends on the number of spots shown on the die. If the die shows n spots, we are to get 2^n dollars. If we play the game just once, obviously the result will "depend on chance." However, if we play many times, we suspect that "things will average out," and the result will be fairly predictable. Let us be definite and replace "many times" by 60,000 times. In 60,000 throws of a die we expect to get about 10,000 "ones"; thus there will be about 10,000 times when we are paid $2. Likewise, there will be about 10,000 times when the die falls "two" and we are paid $4. Similarly, we expect to receive each of the amounts $8, $16, $32, and $64 about 10,000 times. To find the average amount we get we add all the amounts we receive and divide by the number of times we play. We are adding approximately 10,000 of each of the numbers 2, 4, 8, 16, 32, and 64. The 10,000 twos total 20,000; the 10,000 fours total 40,000; etc. Thus, the total amount we expect to receive is $20,000 + 40,000 + 80,000 + 160,000 + 320,000 + 640,000 = 1,260,000$; we receive a total of about $1,260,000. Dividing by the number of games, namely 60,000, we have an average of $1,260,000/60,000 = 21$ dollars per game. To summarize, we expect to receive about $21 per game if we play 60,000 games. It is clear that we would have obtained the same answer, $21, with any number of games; of course, our reasoning did use the fact that 60,000 was "large." Since we pay $20 and expect to receive an average of $21, it is clear that the game is to our advantage. We expect to gain $1 per game, on the average. Very soon we shall define $21 to be the "expected value," also called the "average value," of the random variable that is the amount we are paid.

Now we repeat the computation of the last paragraph in a general form. Let N be the number of times we do the experiment. On the basis of the intuitive ideas with which we started our study of probability, we believe that each point u of Ω will occur about $NP(u)$ times in the N trials. Let us suppose that each u does occur $NP(u)$ times and compute the average of the values assumed by X on that basis. For each $u \in \Omega$, we have $NP(u)$ occurrences of the value $X(u)$. Thus the total of the values assumed by X would be the sum of all the numbers $NP(u)X(u)$ for all $u \in \Omega$. Accordingly the average of the values taken by X would be expected to be this sum divided by N; in other words, it would be expected to be the sum of $P(u)X(u)$ for all $u \in \Omega$. Note that we have not proved anything; all we have done is to find a certain sum that seems to be worth further study.

We define the expected value of a random variable as follows: For each u in Ω, compute the number $P(u)X(u)$. The sum of all these numbers is called the *expected value* of X and is denoted by $\mathbf{E}(X)$. The definition we just gave for $\mathbf{E}(X)$ is expressed in symbols by the following equation:

$$\mathbf{E}(X) = \sum_{u \in \Omega} P(u)X(u).$$

Sometimes we shall refer to the expected value of X as the average value of X or as the mean value of X.

As we said back in Chapter One, the idea of an expected value appears in the correspondence between Fermat and Pascal, at the very beginning of the formal study of probability theory. However, the formal definition of a random variable had to wait until probability theory was formalized in the twentieth century. At that time our definition of an expected value became meaningful. The basic theorem,

$$\mathbf{E}(X) = \sum_{t} tP(X = t),$$

which we shall discuss shortly, is essentially the content of Huygens's third proposition, but the idea behind this formula appears in the Fermat–Pascal correspondence.

Curiously, the important point to make here is about terminology. The word "expectation," which is just about the same in Latin as in English, was used by Huygens in connection with the number we have denoted by $E(X)$. However, the abstract concept of a random variable X not yet existing, the reference was originally to "my expectation." In all of the early works on probability, it was always a person who had an "expectation" of receiving something. More recently, and, it seems to the author, illogically, many books on probability refer to $E(X)$ as the expectation of X. Be that as it may, the reader should recognize that "expected value," "expectation," "mean," and "average" are all synonyms.

We next consider an example. Suppose three coins are tossed. In this first example, we shall write out all the details in full. Let $\Omega = \{hhh, hht, hth, htt, thh, tht, tth, ttt\}$, where the notation is obvious. Let X be the number of coins that fall heads. Then

$$P(hhh) = 1/8, \quad X(hhh) = 3$$
$$P(hht) = 1/8, \quad X(hht) = 2$$
$$P(hth) = 1/8, \quad X(hth) = 2$$
$$P(htt) = 1/8, \quad X(htt) = 1$$
$$P(thh) = 1/8, \quad X(thh) = 2$$
$$P(tht) = 1/8, \quad X(tht) = 1$$
$$P(tth) = 1/8, \quad X(tth) = 1$$
$$P(ttt) = 1/8, \quad X(ttt) = 0.$$

Thus $E(X) = \frac{1}{8} \cdot 3 + \frac{1}{8} \cdot 2 + \frac{1}{8} \cdot 2 + \frac{1}{8} \cdot 1 + \frac{1}{8} \cdot 2 + \frac{1}{8} \cdot 1 + \frac{1}{8} \cdot 1 + \frac{1}{8} \cdot 0$. Hence, $E(X) = 3/2$. We say that we expect $3/2$ heads. Of course, we know that on any one toss of three coins we must get a whole number of heads, not $3/2$. But, *on the average*, we expect $3/2$ heads. Note that now we only have a suspicion that in many tosses of the three coins we would average approximately $3/2$ heads per toss; later on we shall state and prove a theorem along these lines.

The method just used to find $E(X)$ requires adding up one number for each point of Ω; that might be very many numbers. So we look for shortcuts. In the example, we could have combined all of the terms for each value of $X(u)$ into a single term. Thus, instead of considering hht, hth, and thh separately, we could have written

(3/8)2, where 3/8 is the probability that two of the coins fall heads. The computation of $\mathbf{E}(X)$ then would be

$$\mathbf{E}(X) = \frac{1}{8} \cdot 3 + \frac{3}{8} \cdot 2 + \frac{3}{8} \cdot 1 + \frac{1}{8} \cdot 0 = \frac{3}{2}.$$

Now let us consider another example. In a lottery, one million tickets are sold. One ticket—of course, no one knows which ticket in advance—entitles the holder to the first prize, which is \$100,000. Ten tickets win second prizes of \$10,000 each. Ten thousand tickets win \$5 each, as consolation prizes. What is the mean value of a randomly chosen ticket? The obvious sample space contains one million points, one for each ticket. There is no reason to add up one million numbers, even though a computer could do it. Instead, we reason as follows. Only 10,011 tickets win anything. The other 989,989 tickets win \$0. One term of zero is more than enough; we don't need 989,989 such terms. Now consider the consolation prizes. Instead of using 10,000 terms of (.000001)5, we can use one term of (10000)(.000001)5 = (.01)5 = .05. Note that .01 is the total probability of all the sample points corresponding to consolation prizes. Thus .01 is the probability of winning a consolation prize. Now consider the second prizes: ten terms of (.000001)(10000) sum to (.00001)(10000) = .1. Here .00001 is the probability of winning a second prize. The one first prize contributes (.000001)(100000) = .1; note that .000001 is the probability of winning \$100,000. Let X be the value of a randomly chosen ticket. Using our work so far, we have

$$\mathbf{E}(X) = .05 + .1 + .1 = .25.$$

Let us develop the procedure of the last paragraph into a general method. We consider in turn each of the values X can take. Let t be one of these values. The definition of $\mathbf{E}(X)$ involved a term for each $u \in \Omega$ such that $X(u) = t$; these terms were of the form $tP(u)$. The total of all these terms is then $tP(X = t)$. Now we need merely add all these totals, for all values t that X can take, to get $\mathbf{E}(X)$. We write

$$\mathbf{E}(X) = \sum_t tP(X = t).$$

It appears that we have a term in the sum for every number t. But if $P(X = t)$ is zero, which will be the case for almost all values of t, the term is zero, and we simply ignore the term. Thus what the last equation really says is that we may find $\mathbf{E}(X)$ as follows: For each number t such that $P(X = t) \neq 0$, multiply t by $P(X = t)$ and then add up all the products.

Let us consider another example of the use of this formula. A coin is tossed ten times. John pays Tom \$1 if the first toss is heads, \$2 if heads occurs for the first time on the second toss, \$4 if heads occurs for the first time on the third toss, etc. In other words, for each of $n = 1, 2, \ldots, 10$, if heads occurs for the first time of the nth toss, John pays Tom 2^{n-1} dollars. If, however, heads never is thrown, that is, if all ten tosses result in tails, Tom must pay John \$5,000. We seek to find out which player has an advantage, and how big this advantage is. Let X be the net amount John pays Tom; in particular, if tails does come up all ten times, X would be -5000. $\mathbf{E}(X)$ is computed as follows: $P(X = 1) = 1/2$, since John pays \$1 exactly when the first throw is heads; $1(1/2) = 1/2$. The probability that \$2 is paid is $(1/2)(1/2) = 1/4$; $2(1/4) = 1/2$. Likewise, $tP(X = t) = 1/2$ for each of $t = 4, 8, \ldots, 1024$. The remaining value t for which $P(X = t) \neq 0$ is $t = -5000$; $P(X = -5000) = 1/2^{10}$. Thus $\mathbf{E}(X)$ is the sum of $(-5000)(1/2^{10})$ and ten terms each equal to $1/2$. In short,

$$\mathbf{E}(X) = 10(1/2) - 5000/2^{10} = 5 - 5000/1024 = 15/128.$$

This is approximately 12 cents. On the average, John pays Tom about 12 cents per game. After a thousand games, Tom would tend to be approximately \$120 ahead. As we said before, we shall make a precise statement about this later. At that time, we shall see that, since $\mathbf{E}(X)$ is positive, the limit of the probability of Tom being an overall loser after n games as n tends to infinity is zero.

We would not call the game just described fair, because Tom averages a net gain of 12 cents per game. Let us take a more positive point of view. We do call a game fair if the expected value of the amount each player gains is zero. (In the case where there are only two players and one player's loss is the other player's gain, it is clear that the game is fair if we know that one player has an expected gain of zero.) In a fair game, *on the average*, nobody wins. However, it is not obvious what the implications of that statement are. In this and subsequent chapters, we shall explore what the consequences of a game's being fair are.

A special kind of function is a constant function. It is possible that $X(u)$ is the same for all points u in the sample space. For example, suppose some dice are thrown and then some coins are tossed. Then Karen pays Susan $10. Let X be the amount that Karen pays Susan; then we can, if we wish, talk about $\mathbf{E}(X)$. We do talk about $\mathbf{E}(X)$ just to avoid having an exception to our theory. Just as a formality, we compute $\mathbf{E}(X)$: $P(X = t) \neq 0$ only if $t = 10$; $10P(X = 10) = 10$; thus $\mathbf{E}(X) = 10$. We may as well generalize. If $X(u) = c$ for every $u \in \Omega$, then $\mathbf{E}(X) = c$.

We shall next be considering situations that involve more than one random variable. Keep in mind that all our theory involves just one sample space; we consider many examples, but at any one time we have a single sample space. Thus, when we discuss random variables X and Y together, it is unnecessary to say, "on the same sample space"; that is automatically the case.

Let X and Y be random variables. Then $X + Y$ is defined in the obvious way; explicitly, $X + Y$ is that random variable that assigns to each $u \in \Omega$ the sum of the numbers assigned by X and Y. Thus

$$(X + Y)(u) = X(u) + Y(u) \qquad \text{for all } u \in \Omega.$$

Next we define $X - Y$ and XY in a similar manner. Thus we have

$$(X - Y)(u) = X(u) - Y(u) \qquad \text{for all } u \in \Omega;$$
$$(XY)(u) = X(u)\,Y(u) \qquad \text{for all } u \in \Omega.$$

Theorem   *Let X and Y be random variables. Then*

$$\mathbf{E}(X + Y) = \mathbf{E}(X) + \mathbf{E}(Y).$$

Proof   We have $(X + Y)(u) = X(u) + Y(u)$ for all $u \in \Omega$. Thus,

$$(X + Y)(u)P(u) = X(u)P(u) + Y(u)P(u)$$

for all $u \in \Omega$. Adding these equations for all $u \in \Omega$, we have

$$\mathbf{E}(X + Y) = \sum_{u \in \Omega}(X + Y)(u)P(u) = \sum_{u \in \Omega}X(u)P(u) + \sum_{u \in \Omega}Y(u)P(u)$$
$$= \mathbf{E}(X) + \mathbf{E}(Y). \qquad \qquad \square$$

Corollary   *Let $X_1, X_2, \ldots, X_n$ be random variables. Then*

$$\mathbf{E}(X_1 + \cdots + X_n) = \mathbf{E}(X_1) + \cdots + \mathbf{E}(X_n).$$

Proof By repeated applications of the theorem, we have

$$\mathbf{E}(X_1) + \cdots + \mathbf{E}(X_n) = \mathbf{E}(X_1 + X_2) + \mathbf{E}(X_3) + \cdots + \mathbf{E}(X_n)$$
$$= \mathbf{E}(X_1 + X_2 + X_3) + \mathbf{E}(X_4) + \cdots + \mathbf{E}(X_n)$$
$$\vdots$$
$$= \mathbf{E}(X_1 + \cdots + X_{n-1}) + \mathbf{E}(X_n)$$
$$= \mathbf{E}(X_1 + \cdots + X_n).$$

$\square$

Theorem *Let X be a random variable and t be a number. Then*

$$\mathbf{E}(tX) = t\mathbf{E}(X).$$

Proof tX is that random variable such that $(tX)(u) = tX(u)$ for all $u \in \Omega$. Thus $(tX)(u)P(u) = tX(u)P(u)$ for all $u \in \Omega$. Adding all these last equations, we have

$$\mathbf{E}(tX) = \sum_{u \in \Omega}(tX)(u)P(u) = \sum_{u \in \Omega} tX(u)P(u) = t\sum_{u \in \Omega} X(u)P(u) = t\mathbf{E}(X).$$

$\square$

Obviously, the single number $\mathbf{E}(X)$, while important, gives a very incomplete picture of the random variable X. It is remarkable just how much information about X can be given by $\mathbf{E}(X)$ and just one other number, $\mathbf{Var}(X)$, which will be defined in a moment. On the other hand, two numbers can only do so much. Knowing $\mathbf{E}(X)$ and $\mathbf{Var}(X)$, we know a lot, but by no means everything, about X. $\mathbf{E}(X)$ tells us the average of the values we would get for X by doing our experiment many times—at least, this average will "probably" be "close" to $\mathbf{E}(X)$. What we would most like to know in addition is how far from $\mathbf{E}(X)$ we are likely to find X on any one performance of the experiment. In repeating the experiment, do we always get values that are far from $\mathbf{E}(X)$, or do we always come close to $\mathbf{E}(X)$? Quite possibly, sometimes far and sometimes near. For brevity, we set $m = \mathbf{E}(X)$. Now we may regard m as a constant random variable. On this basis, $X - m$ makes sense. In fact, it makes mathematical and practical sense to ask how big $(X - m)^2$ is on the average. Why square? Since we are concerned with how far X is from m, that is, with $|X - m|$, we want to get rid of the sign of $X - m$; 97 is just as far from 100 as 103 is. Clearly $(X - m)^2$ depends only on $|X - m|$. Why not use, for example, $(X - m)^4$ or $|X - m|$ itself? It will turn out, although it is not obvious at this point, that studying $(X - m)^2$ gives us a particularly nice theory. Therefore we will study $(X - m)^2$.

We denote $\mathbf{E}([X - \mathbf{E}(X)]^2)$ by $\mathbf{Var}(X)$. We call $\mathbf{Var}(X)$ the *variance* of X. The reasons for studying the variance of X were discussed in the last paragraph: $\mathbf{Var}(X)$ tells us the extent to which X fluctuates from one performance of the experiment to another.

Let us illustrate with an example. Moe and Joe have an agreement whereby Moe will pay Joe an amount determined by the throw of a die. In detail, the amount to be paid is described in the first two columns of the table below: If u is the number shown on the die, then $X(u)$ is the amount paid.

u	$X(u)$	$Y(u)$	$X(u) - 10$	$Y(u) - 10$	$[X(u) - 10]^2$	$[Y(u) - 10]^2$
1	7	−2990	−3	−3000	9	9000000
2	8	−1990	−2	−2000	4	4000000
3	9	−990	−1	−1000	1	1000000
4	11	1010	1	1000	1	1000000
5	12	2010	2	2000	4	4000000
6	13	3010	3	3000	9	9000000

Using the table, it is easy to find the expected value of the random variable X. We have

$$\mathbf{E}(X) = \frac{1}{6} \cdot 7 + \frac{1}{6} \cdot 8 + \frac{1}{6} \cdot 9 + \frac{1}{6} \cdot 11 + \frac{1}{6} \cdot 12 + \frac{1}{6} \cdot 13 = 10.$$

We contrast Moe and Joe with Sue and Pru. Sue and Pru have an agreement similar to that of Moe and Joe. But the amount Y that Sue pays Pru is described by the first and third columns of the table. For example, if the die falls "two," Sue pays Pru $\$-1990$; in other words, in that case, Pru pays Sue $\$1990$. We compute

$$\mathbf{E}(Y) = \frac{1}{6}(-2990) + \frac{1}{6}(-1990) + \frac{1}{6}(-990) + \frac{1}{6}(1010)$$
$$+ \frac{1}{6}(2010) + \frac{1}{6}(3010)$$
$$= 10.$$

Thus, $\mathbf{E}(X) = \mathbf{E}(Y)$. The difference between X and Y is revealed only when we examine their variances.

The next step is to compute **Var**(X) and **Var**(Y). The beginning of the computation is shown in the fourth, fifth, sixth, and seventh columns of the table above. We finish finding **Var**(X) by finding $\mathbf{E}([X - \mathbf{E}(X)]^2)$; we have

$$\mathbf{Var}(X) = \mathbf{E}([X - \mathbf{E}(X)]^2) = \frac{1}{6} \cdot 9 + \frac{1}{6} \cdot 4 + \frac{1}{6} \cdot 1 + \frac{1}{6} \cdot 1 + \frac{1}{6} \cdot 4 + \frac{1}{6} \cdot 9 = \frac{14}{3}.$$

Similarly, we have

$$\mathbf{Var}(Y) = \mathbf{E}([Y - \mathbf{E}(Y)]^2)$$

$$= \frac{1}{6} \cdot 9 \cdot 10^6 + \frac{1}{6} \cdot 4 \cdot 10^6 + \frac{1}{6} \cdot 1 \cdot 10^6 + \frac{1}{6} \cdot 1 \cdot 10^6$$

$$+ \frac{1}{6} \cdot 10^9 + \frac{1}{6} \cdot 4 \cdot 10^6 + \frac{1}{6} \cdot 9 \cdot 10^6$$

$$= \frac{14}{3} \cdot 10^6.$$

In short, the variance of Y is one million times the variance of X.

Let us try briefly, and informally, to understand the significance to Moe, Joe, Sue, and Pru of the million-fold difference in the variance. As for Moe and Joe, Moe necessarily pays Joe approximately $10. The amount paid may be a few dollars off from $10, but there is no reason for excitement. As for Sue and Pru, one of them will pay the other at least $1000, in round numbers. They probably will be concerned about who pays whom, with several thousand dollars involved. We may say that how Sue and Pru fare is more affected by chance than the fate of Moe and Joe. In a way, the variance of a random variable tells us how large a role chance plays in determining the value of the variable.

If $Y(u) \geq 0$ for all $u \in \Omega$, then it is clear from the definition of $\mathbf{E}(Y)$ that $\mathbf{E}(Y) \geq 0$. Applying this remark to $Y = [X - \mathbf{E}(X)]^2$, we see that $\mathbf{Var}(X) = \mathbf{E}(Y) \geq 0$ always; $\mathbf{Var}(X)$ is *never* negative.

We call the square root of $\mathbf{Var}(X)$ the *standard deviation* of X. [We just saw that $\mathbf{Var}(X)$ is never negative and thus always has a square root.] The name "standard deviation" is used because, as we shall see later, the number so called is a standard for measuring deviations from $\mathbf{E}(X)$.

The following theorem contains a formula for $\mathbf{Var}(X)$ that is usually more convenient in specific examples than the definition of the variance.

Theorem *Let X be a random variable. Then*

$$\mathbf{Var}(X) = \mathbf{E}(X^2) - [\mathbf{E}(X)]^2.$$

Proof To simplify notation, let $m = \mathbf{E}(X)$. By definition, we have $\mathbf{Var}(X) = \mathbf{E}((X - m)^2)$. Thus we have

$$
\begin{aligned}
\mathbf{Var}(X) &= \mathbf{E}(X^2 - 2Xm + m^2) \\
&= \mathbf{E}(X^2) + \mathbf{E}(-2mX) + \mathbf{E}(m^2) \\
&= \mathbf{E}(X^2) - 2m\mathbf{E}(X) + m^2 \\
&= \mathbf{E}(X^2) - 2m^2 + m^2 \\
&= \mathbf{E}(X^2) - m^2,
\end{aligned}
$$

as required. □

We next work a simple problem two ways, first directly and then using the formula just derived. As we shall see, in a simple problem it makes little difference which method we use. In a more complicated situation, the advantage of the new formula is more substantial. To get to the point: Suppose a die is thrown. Bob pays Ray $2 if the die falls "one," "two," or "three." And Bob pays Ray $3 if the die falls "four" or "five." But if the die falls "six," Bob pays Ray $600. In seeking **Var**(X), we must first find **E**(X). We have

$$\mathbf{E}(X) = \frac{1}{2} \cdot 2 + \frac{1}{3} \cdot 3 + \frac{1}{6} \cdot 600 = 102.$$

Directly from the definition of **Var**(X) we have

$$\mathbf{Var}(X) = \frac{1}{2}(-100)^2 + \frac{1}{3}(-99) + \frac{1}{6}(498)^2 = 49601.$$

With easier arithmetic we can find

$$\mathbf{E}(X^2) = \frac{1}{2} \cdot 2^2 + \frac{1}{3} \cdot 3^2 + \frac{1}{6} \cdot 600^2 = 60005.$$

Thus

$$\mathbf{Var}(X) = 60005 - 102^2 = 49601.$$

We end the section with a theorem that gives us a written record of two fairly obvious formulas.

Theorem *Let X be a random variable and t be a number. Then*

$$\mathbf{E}(X + t) = \mathbf{E}(X) + t,$$
$$\mathbf{Var}(X + t) = \mathbf{Var}(X).$$

Proof In the expression $X + t$, the t denotes a constant random variable. We already have noted that $\mathbf{E}(t) = t$. Thus we have $\mathbf{E}(X+t) = \mathbf{E}(X) + \mathbf{E}(t) = \mathbf{E}(X) + t$. Setting $Y = X + t$, we have $Y - \mathbf{E}(Y) = X+t-[\mathbf{E}(X)+t] = X - \mathbf{E}(X)$. Thus $\mathbf{Var}(X+t) = \mathbf{Var}(Y) = \mathbf{Var}(X)$. $\square$

Exercises

1. John provides and tosses a dime and a half-dollar. Richard gets to keep whichever coins fall heads.

   a. Find the expected value of the amount Richard gets.

   b. How much should Richard pay John in advance to make the game fair?

2. Fred tosses two coins. If both fall heads, he wins $10. If just one falls heads, he wins $4. But if both coins fall tails, he must pay a $2 penalty. Find the expected value of Fred's gain.

3. In certain lottery, 5,000,000 tickets are sold for $1 each.

   > 1 ticket wins a prize of $1,000,000.
   > 10 tickets win prizes of $100,000 each.
   > 100 tickets win prizes of $1,000 each.
   > 10,000 tickets win prizes of $10 each.
   > 1,000,000 tickets receive a refund of the purchase price.

   a. Find the mean value of the amount a ticket gets.

   b. Find the expected net gain for each ticket.

4. Six coins are tossed. Alice pays Betty according to the following table:

If no heads	$200
If 1 head	$50
If 2 heads	$10
If 3 heads	$5
If 4 heads	$20

If 5 heads $25

If 6 heads $80

If X is the amount Alice pays, find $\mathbf{E}(X)$.

5. Alan tosses a coin 20 times. Bob pays Alan $1 if the first toss falls heads, $2 if the first toss falls tails and the second heads, $4 if the first two tosses both fall tails and the third heads, $8 if the first three tosses fall tails and the fourth heads, etc. If the game is to be fair, how much should Alan pay Bob for the right to play the game?

6. Find the mean and variance of each of the random variables described below; each of parts a–o refers to a different random variable.

   a. $P(X = -1) = 1/4$, $P(X = 0) = 1/2$, $P(X = 1) = 1/4$.

   b. $P(X = -1) = 1/4$, $P(X = 0) = 1/2$, $P(X = 5) = 1/4$.

   c. $P(X = -5) = 1/4$, $P(X = 0) = 1/2$, $P(X = 5) = 1/4$.

   d. $P(X = -5) = .01$, $P(X = 0) = .98$, $P(X = 5) = .01$.

   e. $P(X = -50) = .0001$, $P(X = 0) = .9998$, $P(X = 50) = .0001$.

   f. $P(X = 1) = 1$.

   g. $P(X = 0) = 1/2$, $P(X = 2) = 1/2$.

   h. $P(X = .01) = .01$, $P(X = 1.01) = .99$.

   i. $P(X = 0) = .99$, $P(X = 100) = .01$.

   j. $P(X = 0) = .999999$, $P(X = 1000000) = .000001$.

   k. $P(X = 0) = 1/2$, $P(X = 2) = 1/2$.

   l. $P(X = 0) = 3/5$, $P(X = 2) = 1/5$, $P(X = 3) = 1/5$.

   m. $P(X = 0) = 4/7$, $P(X = 2) = 2/7$, $P(X = 3) = 1/7$.

   n. $P(X = 0) = 5/8$, $P(X = 2) = 1/8$, $P(X = 3) = 1/4$.

   o. $P(X = 0) = 2/3$, $P(X = 3) = 1/3$.

7. A coin is tossed repeatedly until heads has occurred twice or tails has occurred twice, whichever comes first. Let X be the number of times the coin is tossed. Find:

   a. $\mathbf{E}(X)$.

   b. $\mathbf{Var}(X)$.

8. Suppose that the random variable X takes only the values 0, 2, and 3. Suppose also $\mathbf{E}(X) = 1$. Show that $1 \leq \mathbf{Var}(X) \leq 2$.

9. Suppose $P(X = a) = P(Y = a) = 0$ unless a is one of three given numbers. Suppose also X and Y have the same mean and the same variance. Show that $P(X = a) = P(Y = a)$ for all numbers a.

10. Arthur and Ben play a game as follows: Arthur tosses three Anthony dollars, which he provides. Ben gets to keep these three coins, but Ben must give Arthur a five-dollar bill if all three coins fall heads.

   a. Find the mean and variance of Arthur's net gain.

   b. Find the mean and variance of Ben's net gain.

11. In a game, a die is thrown. Alan pays Betty \$2 if the die falls "one," "two," or "three"; \$3 if it falls "four" or "five"; and \$6 if it falls "six." Let X be the amount Betty receives. Find:

   a. $\mathbf{E}(X)$.

   b. $\mathbf{Var}(X)$.

12. A store has in stock a supply of packages of candy corn. Specifically, suppose there are

   20 packages of 100 candy corns each,
   40 packages of 200 candy corns each,
   20 packages of 300 candy corns each.

   Let X be the number of candy corns in a randomly chosen package. Find:

   a. The mean of X.

   b. The variance of X.

13. Three coins are tossed together. If all three fall heads, they are tossed again. If again all three fall heads, a third toss is made. Continuing in this way, the process goes on until the coins fall some way other than "three heads." Ms. Payor pays Mr. Gettor 2^k dollars, where k is the number of times the coins are tossed. Find the expected value and variance of the amount Mr. Gettor gets. (Note that this exercise is an exception to our current rule

that Ω may contain only finitely many points. Use your best judgement as to how to proceed.)

14. A box contains one twenty-dollar bill and four one-dollar bills. Two bills are randomly drawn, one at a time, without replacement.

   a. Find the expected value of the bill drawn first.

   b. Find the expected value of the total amount drawn.

15. A hat contains two tickets each marked $2, one ticket marked $4, and one ticket marked $20. Mr. Smith draws a ticket and keeps it. Then Ms. Jones draws a ticket. Each receives the number of dollars stated on the ticket. Let X be the amount Mr. Smith gets and Y the amount Ms. Jones gets. Find:

   a. $E(X)$.

   b. $E(Y)$.

   c. $Var(X)$.

   d. $Var(Y)$.

16. A hat contains one thousand-dollar bill and four five-dollar bills. Five persons, one at a time, each draw a bill at random and keep it. Let X_1 be the value of the first person's bill, X_2 the value of the second person's bill, etc. Find:

   a. $E(X_1)$.

   b. $E(X_5)$.

   c. $Var(X_1)$.

   d. $Var(X_3)$.

   e. $E(X_1 + \cdots + X_5)$.

   f. $Var(X_1 + \cdots + X_5)$.

4.2 Computation of Expected Value and Variance

Some basic properties of random variables were introduced in the last section. Now we consider the computation of means and variances in more complicated circumstances. We first develop some formulas that apply in connection with Bernoulli trials. Then we point out that the same technique that derived the formulas also works under more general conditions. But we need just a little more theory, before we get to the problems.

We have seen that the equation

$$\mathbf{E}(X + Y) = \mathbf{E}(X) + \mathbf{E}(Y)$$

holds for any two random variables. The situation is different for the superficially similar equations

$$\mathbf{E}(XY) = \mathbf{E}(X)\mathbf{E}(Y)$$

and

$$\mathbf{Var}(X + Y) = \mathbf{Var}(X) + \mathbf{Var}(Y).$$

These latter equations are not always correct. We next investigate circumstances in which, as we shall see later, they do hold. The definition that comes next describes a condition that is *more* than enough to ensure that both equations hold.

We call random variables $X_1, \ldots, X_n$ *independent* if for every choice of numbers $t_1, \ldots, t_n$ the following condition holds: For each i, let A_i be the event consisting of those $u \subset \Omega$ such that $X_i(u) - t_i$; then the events $A_1, \ldots, A_n$ are independent. It will turn out that all we really need to know is when two random variables are independent. Since it is easy to say when two events are independent, we can rephrase the definition just given into a simpler form when only two random variables are involved. Let X and Y be random variables. Then X and Y are independent if and only if

$$P(X = s, Y = t) = P(X = s)P(Y = t)$$

for every choice of numbers s and t.

The next step is to state and prove a certain lemma. To make the lemma easier to understand, we first consider an example. Five coins are tossed. Let X be the absolute value of the difference between the number of coins that fall heads and the number that fall tails. We make the following computation:

No. of heads	Probability	Value of X	Probability times value of X
0	1/32	5	5/32
1	5/32	3	15/32
2	10/32	1	10/32
3	10/32	1	10/32
4	5/32	3	15/32
5	1/32	5	5/32
			Total 60/32

Is the number we found, 60/32, equal to $\mathbf{E}(X)$? Yes, but why? We assume that we are using the obvious sample space with 32 points, one for each way the coins can fall. Since we added only six terms, we certainly didn't use the definition of $\mathbf{E}(X)$. But, for example, $P(X = 1) = 20/32$ wasn't used either. Thus we didn't use the formula

$$\mathbf{E}(X) = \sum_t tP(X = t).$$

We are somewhere in between. We may justify our work by the following lemma:

Lemma   Let $A_1, \ldots, A_n$ be a partition of Ω and $t_1, \ldots, t_n$ be numbers. Suppose, for each i and each $u \in A_i$, $X(u) = t_i$. Then

$$\mathbf{E}(X) = \sum_{i=1}^{n} t_i P(A_i).$$

Proof   Consider any one of the sets A_i. Then

$$P(A_i) = \sum_{u \in A_i} P(u).$$

Thus

$$t_i P(A_i) = \sum_{u \in A_i} t_i P(u) = \sum_{u \in A_i} X(u)P(u),$$

since $t_i = X(u)$ for each $u \in A_i$. Now we simply add these equations for all i. Since each u belongs to just one A_i, the sum of the right-hand sides is just

$$\sum_{u \in \Omega} X(u)P(u),$$

which by definition is $\mathbf{E}(X)$; hence the conclusion is clear.   □

Theorem *Let X and Y be independent random variables. Then*

$$\mathbf{E}(XY) = \mathbf{E}(X)\mathbf{E}(Y).$$

Proof Let $t_1, \ldots, t_n$ be all the numbers r for which $P(X = r) \neq 0$; let $s_1, \ldots, s_m$ be the numbers r for which $P(Y = r) \neq 0$. Then

$$\mathbf{E}(X) = t_1 P(X = t_1) + \cdots + t_n P(X = t_n),$$
$$\mathbf{E}(Y) = s_1 P(Y = s_1) + \cdots + s_m P(Y = s_m).$$

If we multiple together the sums on the right in these two equations, we get a sum of terms of the form

$$t_i s_j P(X = t_i) P(Y = x_j);$$

in this last sum, we have one such term for each choice of i and j. Since X and Y are independent, $P(X = t_i)P(Y = s_j) = P(X = t_i, Y = s_j)$. Let B_{ij} consist of those $u \in \Omega$ such that $X(u) = t_i$ and $Y(u) = s_j$. We have just seen that $\mathbf{E}(X)\mathbf{E}(Y)$ is the sum of $t_i s_j P(B_{ij})$ over all choices of i and j. Given $u \in \Omega$, there is just one i for which $X(u) = t_i$ and just one j for which $Y(u) = s_j$; thus u belongs to just one of the sets B_{ij}. In other words, the sets B_{ij} form a partition of Ω. If $u \in B_{ij}$, then $X(u) = t_i$, $Y(u) = s_j$, and hence $(XY)(u) = t_i s_j$. Thus by the lemma, $\mathbf{E}(XY)$ is the sum of $t_i s_j P(B_{ij})$ over all choices of i and j. Since we have already noted that $\mathbf{E}(X)\mathbf{E}(Y)$ is equal to the same sum, the proof is complete. □

Lemma *Let X and Y be random variables. Suppose $\mathbf{E}(XY) = \mathbf{E}(X)\mathbf{E}(Y)$. Then*

$$\mathbf{Var}(X + Y) = \mathbf{Var}(X) + \mathbf{Var}(Y).$$

Proof By an earlier theorem,

$$\mathbf{Var}(X + Y) = \mathbf{E}((X + Y)^2) - [\mathbf{E}(X + Y)]^2.$$

We have,

$$\mathbf{E}((X + Y)^2) = \mathbf{E}(X^2 + 2XY + Y^2)$$
$$= \mathbf{E}(X^2) + \mathbf{E}(2XY) + \mathbf{E}(Y^2)$$
$$= \mathbf{E}(X^2) + 2\mathbf{E}(X)\mathbf{E}(Y) + \mathbf{E}(Y^2).$$

On the other hand,

$$[E(X + Y)]^2 = [E(X) + E(Y)]^2$$
$$= [E(X)]^2 + 2E(X)E(Y) + [E(Y)]^2.$$

Subtracting $[E(X + Y)]^2$ from $E((X + Y)^2)$, we have

$$\text{Var}(X + Y) = E(X^2) + E(Y^2) - [E(X)]^2 - [E(Y)]^2$$
$$= \text{Var}(X) + \text{Var}(Y).$$

$\square$

Theorem *Suppose* $X_1, \ldots, X_n$ *are independent random variables. Then*

$$\text{Var}(X_1 + \cdots + X_n) = \text{Var}(X_1) + \cdots + \text{Var}(X_n).$$

Proof Since X_i and X_j are independent whenever $i \neq j$, $E(X_iX_j) = E(X_i)E(X_j)$ provided $i \neq j$. Thus we have, for all suitable k,

$$E\left((X_1 + \cdots + X_k)X_{k+1}\right) = E(X_1X_{k+1} + \cdots + X_kX_{k+1})$$
$$= E(X_1)E(X_{k+1}) + \cdots + E(X_k)E(X_{k+1})$$
$$= [E(X_1) + \cdots + E(X_k)]E(X_{k+1})$$
$$= E(X_1 + \cdots + X_k)E(X_{k+1}).$$

By the lemma then,

$$\text{Var}(X_1) + \cdots + \text{Var}(X_n) = \text{Var}(X_1 + X_2) + \text{Var}(X_3) + \cdots + \text{Var}(X_n)$$
$$= \text{Var}(X_1 + X_2 + X_3) + \text{Var}(X_4)$$
$$+ \cdots + \text{Var}(X_n)$$
$$= \text{Var}(X_1 + \cdots + X_{n-1}) + \text{Var}(X_n)$$
$$= \text{Var}(X_1 + \cdots + X_n).$$

$\square$

In the hypothesis of the last theorem, we assumed $X_1, \ldots, X_n$ to be independent. Obviously, we could merely have required each two of $X_1, \ldots, X_n$ to be independent; that is all we used in the proof. We used the first condition for brevity. In most applications, it is obvious that both conditions are satisfied.

We work an example using independent random variables. One hundred dice are thrown; let X be the total number of spots shown on the dice. We seek $E(X)$ and $\text{Var}(X)$. Let X_n be the number of spots shown on the nth die. Then $X = X_1 + \cdots + X_{100}$. We first find $E(X_i)$

and $\mathbf{Var}(X_i)$ for one, and hence all, i. We have

$$\mathbf{E}(X_i) = \frac{1}{6} \cdot 1 + \frac{1}{6} \cdot 2 + \frac{1}{6} \cdot 3 + \frac{1}{6} \cdot 4 + \frac{1}{6} \cdot 5 + \frac{1}{6} \cdot 6 = \frac{7}{2},$$

$$\mathbf{E}(X_i^2) = \frac{1}{6} \cdot 1 + \frac{1}{6} \cdot 4 + \frac{1}{6} \cdot 9 + \frac{1}{6} \cdot 16 + \frac{1}{6} \cdot 25 + \frac{1}{6} \cdot 36 = \frac{91}{6},$$

$$\mathbf{Var}(X_i) = \left(\frac{7}{2}\right)^2 - \frac{91}{6} = \frac{35}{12}.$$

Now we have $\mathbf{E}(X) = \mathbf{E}(X_1) + \cdots + \mathbf{E}(X_{100}) = 100\mathbf{E}(X_i) = 350$ and, using the independence of $X_1, \ldots, X_{100}$, $\mathbf{Var}(X) = \mathbf{Var}(X_1) + \cdots + \mathbf{Var}(X_{100}) = 100\mathbf{Var}(X_i) = 875/3$.

We need just one more theorem in this chapter.

Theorem *Let X be a random variable and t be a number. Then*

$$\mathbf{Var}(tX) = t^2\mathbf{Var}(X).$$

Proof

$$\mathbf{Var}(tX) = \mathbf{E}((tX)^2) - [\mathbf{E}(tX)]^2 = \mathbf{E}(t^2 X^2) - [t\mathbf{E}(X)]^2$$
$$= t^2\mathbf{E}(X^2) - t^2 [\mathbf{E}(X)]^2 = t^2\mathbf{Var}(X).$$

$\square$

We return to the 100 dice thrown just before the statement of the theorem. Let Y be the average number of spots shown on the 100 dice. In other words, let $Y = X/100$. We have $\mathbf{E}(Y) = \mathbf{E}(X/100) = \mathbf{E}(X)/100 = 7/2$. Also, $\mathbf{Var}(Y) = \mathbf{Var}(X/100) = \mathbf{Var}(X)/10000 = 7/240$. Note that the expected value is the same as that for throwing a single die, but the variance is much less than that for a single die. We leave a discussion of the significance of this computation for the next chapter.

Now we are ready to remove the restriction that the sample space contain only finitely many points. Suppose Ω contains infinitely many points. In accordance with the assumption we made in Chapter One, we may designate these points as $u_1, u_2, u_3, \ldots$. We make the obvious definition of $\mathbf{E}(X)$; that is, we let

$$\mathbf{E}(X) = \sum_{i=1}^{\infty} P(u_i)X(u_i).$$

There are some problems: The series may be divergent. In that case, X has no expected value. The choice of which point of Ω is u_1, which

u_2, etc. is arbitrary. If we change the order, we may change the sum of the series. To avoid this difficulty, we agree to define $\mathbf{E}(X)$ only when the series above is absolutely convergent. It is a property of such series that the series remains absolutely convergent with the same sum if the terms are put into a different order. To repeat, if the series above is not absolutely convergent, X has no expected value. If the series is absolutely convergent, we say, "$\mathbf{E}(X)$ exists." With the new definition of $\mathbf{E}(X)$, all the theory we just did goes through with obvious minor modifications. We give a description of the changes necessary in the next paragraph.

We restate the theorems just discussed here partly to include the case of infinite Ω and partly just for the convenience of having them all together. We shall not worry about proofs; the proofs are essentially unchanged from the finite case, except that we must use the properties of infinite series. Note that the definition of variance, $\mathbf{Var}(X) = \mathbf{E}([X - \mathbf{E}(X)]^2)$, still makes sense. If X has no expected value, then it has no variance. If X has an expected value but $[X - \mathbf{E}(X)]^2$ does not, then X still has no variance. Now we list the theorems:

Theorem *Let X and Y be random variables. Suppose $\mathbf{E}(X)$ and $\mathbf{E}(Y)$ exist. Then $\mathbf{E}(X + Y)$ exists and*

$$\mathbf{E}(X + Y) = \mathbf{E}(X) + \mathbf{E}(Y).$$

Theorem *Let X be a random variable and t be a number. Suppose $\mathbf{E}(X)$ exists. Then $\mathbf{E}(tX)$ exists and*

$$\mathbf{E}(tX) = t\mathbf{E}(X).$$

Theorem *Let X be a random variable. Suppose $\mathbf{E}(X)$ and $\mathbf{E}(X^2)$ exist. Then $\mathbf{Var}(X)$ exists and*

$$\mathbf{Var}(X) = \mathbf{E}(X^2) - [\mathbf{E}(X)]^2.$$

Theorem *Let X and Y be independent random variables. Suppose $\mathbf{E}(X)$ and $\mathbf{E}(Y)$ exist. Then $\mathbf{E}(XY)$ exists and*

$$\mathbf{E}(XY) = \mathbf{E}(X)\mathbf{E}(Y).$$

Theorem *Suppose $X_1, \ldots, X_n$ are independent random variables, each having a variance. Then $\mathbf{Var}(X_1 + \cdots + X_n)$ exists and*

$$\mathbf{Var}(X_1 + \cdots + X_n) = \mathbf{Var}(X_1) + \cdots + \mathbf{Var}(X_n).$$

Theorem   *Let X be a random variable and t be a number. Suppose* **Var**(X) *exists. Then* **Var**(tX) *exists and*

$$\mathbf{Var}(tX) = t^2 \mathbf{Var}(X).$$

There is a certain method for finding expected values that often greatly simplifies their computation. We shall illustrate this method in several examples, the first two of which are very important. The method is absurdly simple. Suppose we want to find the expected value of a somewhat complicated random variable X. If we can find much simpler random variables $X_1, \ldots, X_k$ such that $X = X_1 + \cdots + X_k$, we can use the fact that $\mathbf{E}(X) = \mathbf{E}(X_1) + \cdots + \mathbf{E}(X_k)$ to find $\mathbf{E}(X)$. Further details of the method become clear in the examples we work next.

Suppose we are going to make n Bernoulli trials. We let p and q have their usual meanings. Let X be the number of successes we shall get; we seek $\mathbf{E}(X)$. Let

$$X_1 = \begin{cases} 1 & \text{if the first trial results in success,} \\ 0 & \text{otherwise.} \end{cases}$$

For those who like formality, we reword the definition: Let A be the event, "The first trial results in success." We define the random variable X_1 by letting

$$\begin{aligned} X_1(u) &= 1 & \text{for each } u \in A, \\ X_1(u) &= 0 & \text{for each } u \in \overline{A}. \end{aligned}$$

We define $X_2, X_3, \ldots, X_n$ in the same way using the second, third,...,nth trials. Consider $X_1 + \cdots + X_n$. At each point u of the sample space, the value of $X_1(u) + \cdots + X_n(u)$ is a sum of zeros and ones—a one for each trial that results in success when that point is the point that occurs. In short, $X = X_1 + \cdots + X_n$. $\mathbf{E}(X_1)$ is easy to find: $P(X_1 = 1)$ is the probability of success on the first trial; thus, $P(X_1 = 1) = p$. Likewise $P(X_1 = 0) = q$. Thus we have

$$\mathbf{E}(X_1) = 0 \cdot P(X_1 = 0) + 1 \cdot P(X_1 = 1) = p.$$

In exactly the same way we see that $\mathbf{E}(X_2) = p$, $\mathbf{E}(X_3) = p, \ldots, \mathbf{E}(X_n) = p$. Now we have

$$\begin{aligned}
\mathbf{E}(X) &= \mathbf{E}(X_1 + \cdots + X_n) \\
&= \mathbf{E}(X_1) + \cdots + \mathbf{E}(X_n) \\
&= p + \cdots + p \quad (\text{with } n \text{ terms}) \\
&= np.
\end{aligned}$$

This important formula for the expected number of successes in n Bernoulli trials should be memorized.

It is also easy to find the variance of the number of successes in n Bernoulli trials. We continue using the notation of the last paragraph:

$$\mathbf{E}(X_1^2) = 0 \cdot P(X_1 = 0) + 1^2 \cdot P(X_1 = 1) = p.$$

Thus

$$\mathbf{Var}(X_1) = \mathbf{E}(X_1^2) - [\mathbf{E}(X_1)]^2 = p - p^2 = p(1 - p) = pq.$$

Now note that $X_1, \ldots, X_n$ are independent. Thus

$$\begin{aligned}
\mathbf{Var}(X) &= \mathbf{Var}(X_1) + \cdots + \mathbf{Var}(X_n) \\
&= pq + \cdots + pq \quad (\text{with } n \text{ terms}) \\
&= npq.
\end{aligned}$$

Again, this formula should be memorized.

Now consider an obvious alternative experiment. We make Bernoulli trials until we get r successes; we seek to know the average number of trials necessary to get the r successes. We started the discussion of this experiment in the last chapter. Before computing an expected value, we must supply the formal definition of a suitable sample space. We leave it to the reader to decide if our mathematical abstraction fairly models their intuitive ideas.

We first do the case $r = 1$; then we do the general case. When $r = 1$, we are investigating the number of trials necessary to get just one success. Since this number is a positive integer, we may as well let $\Omega = \{1, 2, 3, \ldots\}$. For each $k \in \Omega$, we define $P(k) = q^{k-1}p$. (How do we get that number? It is the probability, determined for a different experiment and sample space, that in k Bernoulli trials success occurs for the first time on the last trial. But here we are doing abstract mathematics and just making a formal definition.)

Next we define

$$P(A) = \sum_{k \in A} P(k)$$

for all subsets A of Ω. We must check that the basic assumptions of Chapter One are satisfied. The only one of them that is not obvious is $P(\Omega) = 1$. We have $P(\Omega) = P(1) + P(2) + \cdots = p + qp + q^2p + \cdots$. We recognize the series as a geometric series with first term $a = p$ and ratio $r = q$. By the well-known formula,

$$S = \frac{a}{1 - r},$$

for the sum of the geometric series $a + ar + ar^2 + ar^3 + \cdots$, we have

$$P(\Omega) = \frac{p}{1 - q} = \frac{p}{q} = 1.$$

(Informally, the computation just completed has the following significance: The probability that we get a success sooner or later has the value one. In other words, we're sure to get a success eventually.) Thus we have established a formal sample space and probability function.

Even in the last paragraph, the author could not force himself to avoid parenthetical intuitive explanations. Having indicated how to be formal if we want to be, let us agree that we prefer to be rather informal. Once the basic ideas of the theory of probability and that of infinite series are mastered, the technical details are easily supplied.

Finally we really do the problem at hand. Recall we are going to make Bernoulli trials until we get a success. Let X be the number of trials necessary. For each of $n = 1, 2, \ldots,$ we define a random variable X_n as follows:

$$X_n = \begin{cases} 1 & \text{if at least } n \text{ trials are necessary} \\ 0 & \text{otherwise.} \end{cases}$$

In other words, if the first $n - 1$ trials all result in failure, then X_n assumes the value 1; if a success occurs before n trials are made, then X_n assumes the value 0. In particular, X_1 is the constant random variable that assigns the number 1 to every point of the sample space; the first trial is always needed. A moment's reflection convinces us that $X = X_1 + X_2 + \cdots$; the reasoning is the same as in the last example. (The ridiculously simple key idea of the method we are using is the following: To count how many times something happens, just add 1 to your running total each time it happens.) We have no proof that $E(X) = E(X_1) + E(X_2) + \cdots$, since the theorem about the expected value of a sum does not apply to infinite sums. Since our X_n are never negative, that equation can be shown to be correct for them. We shall apply the equation here, and we shall give an alternative derivation of the formula for $E(X)$ in Chapter Seven. Let us compute $E(X_n)$. $P(X_n = 1)$ is the probability of failure on all of the first $n - 1$ trials; thus $P(X_n = 1) = q^{n-1}$. (This equation holds even if $n = 1$.) Thus

$$E(X_n) = 1 \cdot P(X_n = 1) + 0 \cdot P(X_n = 0) = q^{n-1}.$$

Hence

$$E(X) = E(X_1) + E(X_2) + \cdots = 1 + q + q^2 + \cdots.$$

Applying the standard formula for the sum of a geometric series, we have

$$E(X) = \frac{1}{1 - q}.$$

Since $1 - q = p$, $E(X) = 1/p$ is the formula we seek. [Since the X_n are not independent, we have no easy way to find $\mathrm{Var}(X)$. We defer the computation of the variance to Chapter Seven.]

Now suppose we seek r successes instead of just one. Let X be the number of Bernoulli trials necessary to get r successes. It is easy to find $E(X)$. Let X_1 be the number of trials necessary to get the first success; let X_2 be the number of *additional* trials necessary to get the second success; and so on up to X_r. Then $X = X_1 + \cdots + X_r$. Hence, $E(X) = E(X_1) + \cdots + E(X_r)$. By the formula we just derived, not only does $E(X_1) = 1/p$, but $E(X_i) = 1/p$ for all i. Thus, $E(X) = r(1/p) = r/p$.

As we said before, the method just used to derive certain formulas is important in itself. We shall give some examples to illustrate that point. But first we should make a record of the formulas by summarizing them in a table. Even before doing that, it will be convenient to introduce some terminology.

An example will help explain what we're going to say. A pair of dice are thrown; let X be the number shown on the first die and Y be the number shown on the second die. Clearly X and Y are very much alike. Viewed separately as abstract mathematical objects, they are essentially identical. However, X and Y are certainly two distinct things. When the experiment is done, X and Y will probably assume different values. We may demonstrate the difference numerically by computing $\mathbf{E}(X^2)$ and $\mathbf{E}(XY)$ and getting two different numbers; the details appear in Exercise 41. In a case like this, we say the two random variables have the same distribution. More precisely, we say two random variables X and Y have the same distribution if

$$P(X = k) = P(Y = k) \qquad \text{for all } k.$$

In some cases, the values of $P(X = k)$ for a random variable X are given by a well-known formula. In such a case, a name may be assigned to the formula. For example, to say X has a binomial distribution is to say there are positive numbers p and q with $p+q = 1$ and a positive integer n such that

$$\begin{aligned} P(X = k) &= p^k q^{n-k} && \text{for } k = 0, 1, 2, \ldots, n, \\ P(X = k) &= 0 && \text{for all other } k. \end{aligned}$$

Whether X is obtained from a system of Bernoulli trials is irrelevant. As long as $P(X = k)$ is given by the formula announced, X has a binomial distribution.

Now we are ready to make a record of the formulas for means and variances we have derived so far. The asterisks in the table below denote values that we have not yet computed; the completed table appears at the end of the book. As we just said, a random variable X is said to have a binomial distribution if $P(X = k)$ has, for all k, the value shown in the table for "binomial"; likewise we speak of a Bernoulli distribution, a geometric distribution, and a Pascal distribution. The table gives means and variances for such random variables.

Now we return to the method used to derive the means and variances in the table. This method is very important in itself. We use it next to work three examples. In thinking about these examples, note particularly that it can be easier to find $\mathbf{E}(X)$ than to find the probabilities of X taking various values.

Example 1

Suppose a committee with six members is to be formed by randomly selecting six of the twelve senators from the New England states.

Name	k for which $P(X = k) \neq 0$	$P(X = k)$ for these k	$\mathbf{E}(X)$	$\mathbf{Var}(X)$	Description
Bernoulli	0,1	q, p	p	pq	One trial; 1 if success, 0 if failure
Binomial	$0, 1, \ldots, n$	$\binom{n}{k}p^k q^{n-k}$	np	npq	Number of successes in n Bernoulli trials
Geometric	$1, 2, 3, \ldots$	$q^{k-1}p$	$\dfrac{1}{p}$	***	Number of Bernoulli trials needed to get one success
Pascal	$r, r+1, \ldots$	$\binom{k-1}{r-1}p^r q^{k-r}$	$\dfrac{r}{p}$	***	Number of Bernoulli trials needed to get r successes

What is the expected number of states to be represented on the committee? Let

$$X_1 = \begin{cases} 1 & \text{if Maine is represented,} \\ 0 & \text{if Maine is not represented.} \end{cases}$$

Likewise, define $X_2, \ldots, X_6$ corresponding to the other five states (in any order). Then, if the number of states represented is denoted by X, we have $X = X_1 + \cdots + X_6$. Since clearly $\mathbf{E}(X_1) = \mathbf{E}(X_2) = \cdots = \mathbf{E}(X_6)$, it follows that $\mathbf{E}(X) = \mathbf{E}(X_1) + \cdots + \mathbf{E}(X_6) = 6\mathbf{E}(X_1)$. To find $\mathbf{E}(X_1)$, we find the probability that Maine is represented. The probability that Maine is *not* represented is

$$\frac{\binom{10}{6}}{\binom{12}{6}} = \frac{5}{22}.$$

Thus $\mathbf{E}(X_1) = 1 \cdot P(X_1 = 1) + 0 \cdot P(X_1 = 0) = P(X_1 = 1) = 1 - 5/22 = 17/22$. We conclude that $\mathbf{E}(X) = 6(17/22) = 51/11$. $\qquad\square$

Example 2

Suppose that an urn contains 100 balls, numbered from 1 to 100. Jane is to draw balls at random, one at a time, without replacement, until she draws a ball with a lower number than one she drew earlier. She will be paid \$1 for each ball she draws. How much does she get—of course we mean, how much does she get on the average?

Now, for each positive integer r, the probability that Jane draws just r balls is not hard to find. But we shall see that we do not need these probabilities; there is an easier way to do the problem. The first step we take is the same whether we find the probabilities or not. We change the description of the procedure Jane follows to provide that she continues to draw balls until all the balls are drawn. However, she still is paid a dollar only for each ball up to and including the first ball that has a lower number than a ball drawn earlier. Thus she receives k or more dollars if and only if the first $k - 1$ balls are drawn in numerical order; the probability of this event is clearly $1/(k - 1)!$. Let X be the amount Jane receives. We define random variables $X_1, \ldots, X_{100}$ as follows:

$$X_k = \begin{cases} 1 & \text{if Jane receives at least } k \text{ dollars,} \\ 0 & \text{if Jane receives no more than } k - 1 \text{ dollars.} \end{cases}$$

In forming $X_1 + X_2 + \cdots + X_{100}$, we add up zeros and ones; we consider how many ones. In those circumstances where Jane receives r dollars, $X_k = 1$ exactly when $k \leq r$; thus in this case there are r ones. It follows that $X = X_1 + \cdots + X_{100}$. Thus $\mathbf{E}(X) = \mathbf{E}(X_1) + \cdots + \mathbf{E}(X_{100})$. We already noted, but stated it a different way, that $P(X_k = 1) = 1/(k - 1)!$. Thus, $\mathbf{E}(X_k) = 1/(k - 1)!$, and hence

$$\mathbf{E}(X) = \frac{1}{0!} + \frac{1}{1!} + \frac{1}{2!} + \cdots + \frac{1}{99!}.$$

$\mathbf{E}(X)$ is thus very, very close to

$$e = 1 + 1 + \frac{1}{2!} + \frac{1}{3!} + \cdots$$

To the nearest cent, Jane receives \$2.72 on the average.

The reasoning just completed would work even if we had considered starting with a different number of balls. We would have obtained

$$\mathbf{E}(X) = 1 + 1 + \frac{1}{2!} + \frac{1}{3!} + \cdots + \frac{1}{(n - 1)!}$$

if we had used n balls. The surprising thing is how little the amount Jane gets depends on the number of balls originally in the urn. The problem really only makes sense with at least three balls; thus we can say $\mathbf{E}(X)$ is always between \$2.50 and \$2.72. It is \$2.72 to the nearest cent if there are at least six balls in all. □

Example 3

Suppose we have 100 letters, each addressed to a different person by name. We prepare an envelope for each letter. Suppose we are called away for a while. In our absence, a helpful, but not very clever, person puts the letters in the envelopes and mails them off. How many people get the correct letter? Obviously, that number is a random variable X. The probabilities of X taking various values are somewhat hard to find; we defer the computation of these probabilities to Chapter Seven. However, $\mathbf{E}(X)$ is very easy to find. Number the letters from 1 to 100. Let

$$X_i = \begin{cases} 1 & \text{if the } i\text{th letter is in the correct envelope,} \\ 0 & \text{otherwise.} \end{cases}$$

Then $X = X_1 + \cdots + X_{100}$; hence, $\mathbf{E}(X) = \mathbf{E}(X_1) + \cdots + \mathbf{E}(X_{100})$. Since clearly $\mathbf{E}(X_1) = \cdots = \mathbf{E}(X_{100})$, we have $\mathbf{E}(X) = 100\mathbf{E}(X_1)$. $P(X_1 = 1)$ is the probability that the first letter gets in the correct one out of 100 envelopes. Thus, $P(X_1 = 1) = 1/100$, and hence $\mathbf{E}(X_1) = 1/100$. Therefore $\mathbf{E}(X) = 1$; on the average, just one person gets the right letter. It is somewhat surprising that the number 100 did not affect the final answer; we would have gotten $\mathbf{E}(X) = 1$ even if we had used any other number in place of 100. This problem is a good example of a situation in which it is easy to find an expected value, but hard to find the probabilities behind it. □

Exercises

17. Suppose X and Y have the same mean and have the same variance. Suppose also X and Y are independent. Then

$$\mathbf{E}((X - Y)^2) = 2\mathbf{Var}(X).$$

18. Suppose X and Y have the same variance. Show that

$$\mathbf{E}\left((X+Y)(X-Y)\right) = \mathbf{E}(X+Y)\mathbf{E}(X-Y).$$

19. Suppose that X and Y are independent and that each has mean 3 and variance 1. Find the mean and variance of $X+Y$ and XY.

20. Suppose X and Y are random variables such that $\mathbf{E}(XY) = 0$. Suppose also each of X and Y has mean 1 and variance 3. Find the variance of $X+Y$.

21. Show that

$$\mathbf{Var}(X+Y) + \mathbf{Var}(X-Y) = 2\mathbf{Var}(X) + 2\mathbf{Var}(Y).$$

22. Four coins are tossed. Let X be the number of heads obtained. Find:

   a. The mean of X.

   b. The standard deviation of X.

23. Find the mean and variance of the total number of spots obtained when 60 dice are thrown.

24. An urn contains five balls numbered 1, 2, 3, 4, 5. A ball is drawn at random; let X be the number shown on the ball. This ball is replaced. Then 200 balls are drawn, one at a time, with replacement. Let Y be the sum of the numbers on the 200 balls. Let Z be the average of the numbers on the 200 balls. Find

   a. $\mathbf{E}(X)$.

   b. $\mathbf{E}(Y)$.

   c. $\mathbf{Var}(X)$.

   d. $\mathbf{Var}(Y)$.

   e. $\mathbf{E}(Z)$.

   f. $\mathbf{Var}(Z)$.

25. 2500 coins are tossed. Let X be the number of heads obtained. Find $\mathbf{E}(X)$, $\mathbf{Var}(X)$, and $\mathbf{E}(X^2)$.

26. Cards are drawn, one at a time, from a standard deck; each card is replaced before the next one is drawn. Let X be the number of draws necessary to get an ace. Find $\mathbf{E}(X)$.

27. Cards are drawn as in the last problem. Let Y be the number of draws necessary so that ten of the draws result in spades. Find $E(Y)$.

28. Suppose that in bobbing for apples there is a probability of 2/3 of getting an apple on any particular try.

    a. Let X be the total number of apples obtained in eight tries. Find the mean and standard deviation of X.

    b. Let Y be the total number of tries it takes to get a total of four apples. Find $E(Y)$.

29. Suppose in a certain game a player receives $1 for each "five" and each "six" thrown on a single die.

    a. If the die is thrown six times, find the mean and variance of the amount the player receives.

    b. If the player throws repeatedly until he gets $10, find the mean of the number of throws necessary.

30. Alice and Betty each provide and toss 100 dimes. Alice keeps all the dimes that fall heads and Betty keeps those that fall tails. Let X be Alice's net gain in cents. Find $E(X)$ and $Var(X)$.

31. A die is thrown and eight coins are tossed. Let X be the number shown on the die and Y be the number of coins that fall heads. Find:

    a. $E(X)$.

    b. $E(Y)$.

    c. $Var(X)$.

    d. $Var(Y)$.

    e. $E(X + Y)$.

    f. $Var(X + Y)$.

    g. $E(XY)$.

    h. $Var(XY)$.

32. A Kennedy half-dollar, an Eisenhower dollar, an Anthony dollar, and a Roosevelt dime are tossed once. Ms. Kear gets to keep all that fall heads. What are the mean and variance of the amount she gets to keep?

33. $3.00 worth of nickels and $2.40 worth of dimes are tossed. Find the mean and variance of the value of the coins that fall heads.

34. Ten dimes, 30 Eisenhower dollars, 20 Anthony dollars, and 20 nickels are tossed. Ms. Dean gets to keep those that fall heads. Let X be the number of coins Ms. Dean gets and Y be their value in cents. Find:
    a. $E(X)$.
    b. $E(Y)$.
    c. $Var(X)$.
    d. $Var(Y)$.
    e. $E(X + Y)$.
    f. $Var(X + Y)$.

35. A bag contains 12 coins as follows: 3 nickels, 4 dimes, and 5 quarters. Four coins are drawn from the bag at random without replacement. Let X be the value of the coins drawn. Find $E(X)$.

36 Assume that 40% of the "public" likes a certain television program. Find the mean and standard deviation of the number of persons in a randomly chosen group of 100 that like the program.

37. A coin is tossed repeatedly until tails appears. Find the mean of the number of tosses necessary.

38. A coin is tossed repeatedly until heads has appeared four times. Find the mean of the number of tosses necessary.

39. A coin is tossed repeatedly until either tails appears or heads has appeared four times, whichever comes first. Find the mean and variance of the number of tosses necessary.

40. Eight United States pennies and four Canadian pennies are tossed. Let X be the number of U.S. pennies and Y the number of Canadian pennies that fall heads. Complete the table below. (Warning: The hard part of this problem is to decide in what order the required items should be computed.)

$E(X) =$	$E(X^2) =$	$Var(X) =$
$E(Y) =$	$E(Y^2) =$	$Var(Y) =$
$E(X + Y) =$	$E((X + Y)^2) =$	$Var(X + Y) =$
$E(XY) =$	$E((XY)^2) =$	$Var(XY) =$

41. A pair of dice is thrown. Let Z_1 be the number shown on the first die and Z_2 be the number shown on the second die. Let $X = Z_1 + Z_2$ and $Y = Z_1 Z_2$. Find

 a. $E(Z_1)$.

 b. $E(Z_2^2)$.

 c. $E(X)$.

 d. $E(Y)$.

 e. $E(XY)$.

 f. $E(X^2)$.

 g. $E(Y^2)$.

42. Four dimes and four nickels are tossed. Let Z_1 be the number of dimes that fall heads. Let Z_2 be the number of nickels that fall heads. Let $X = Z_1 + Z_2$ and $Y = Z_1 Z_2$. For these random variables, find items a–g of the last problem.

43. In a problem involving independent random variables X and Y, a probability student correctly finds that

$$E(Y) = 2, \quad E(X^2 Y) = 6, \quad E(XY^2) = 8, \quad \text{and } E((XY)^2) = 24.$$

 Then a goblin destroys the data the student used. Find $E(X)$.

44. On an imaginary television show, there are five identical closed boxes. One box contains $10,000, two boxes contain $1000 each, one box contains $1, and the last box contains a miniature stop sign. A contestant is allowed to select a box, open it, and keep the contents. The contestant is allowed to repeat this process until the box with the stop sign is opened; at this point, the game ends. Find the mean amount the contestant gets.

45. An urn contains six balls of each of the three colors: red, blue, and green. Find the expected number of different colors obtained when three balls are drawn:

 a. with replacement;

 b. without replacement.

46. A poker hand contains five cards. Find the mean of each of the following:

 a. The number of spades in a poker hand.

b. The number of different suits in a poker hand.

c. The number of aces in a poker hand.

d. The number of different face values in a poker hand.

e.–h. Do a–d for a bridge hand (13 cards).

47. Cards are drawn, one at a time, from a standard deck, without replacement.

   a. Let X be the number of draws necessary to get an ace. Find $E(X)$.

   b. Let X_2, X_3, and X_4 be the number of draws necessary to get two, three, and four aces, respectively. Find $E(X_2)$, $E(X_3)$, and $E(X_4)$.

48. In a well-shuffled deck of cards, what is the expected number of spades between the ace of diamonds and the king of hearts?

49. A purse contains 12 quarters and 2 pennies. All the coins are to be drawn, one at a time, without replacement. You are to keep all the quarters that are drawn between the two pennies. What is your expectation?

50. a. A purse contains $101 in Anthony dollars and $1 in pennies. Smith is allowed to draw and keep coins, one at a time, until he has drawn 30 pennies. Find the expected value of the value of the coins he keeps.

   b. Suppose the value of each coin drawn is credited to Smith, and then the coin is replaced before the next coin is drawn. Now what is the expected value of the amount credited to Smith?

51. Four letters are drawn, one at a time, from

   WALLA WALLA.

   What is the expected number of different letters to be drawn?

52. An urn contains 5 red balls and 15 blue balls. The balls are drawn, one at a time, without replacement, until three red balls have been drawn. Find the expected number of balls to be drawn.

53. Repeat the last exercise assuming the balls are drawn with replacement.

54. A committee is to consist of 50 randomly chosen United States senators. Find the expected number of different states to be represented on the committee.

55. We choose letters, one at a time, at random, from the word

    CHOOSE

    until the C is obtained. What is the expected number of choices necessary if the letters are chosen:

    a. With replacement?

    b. Without replacement?

56. We randomly choose letters, one at a time, without replacement, from the word

    CHOOSE

    until both Os have been obtained. What is the expected number of choices necessary?

57. We randomly choose letters, one at a time, without replacement, from the word

    DEPOSITOR.

    What is the expected number of choices if:

    a. We stop when any one letter of the word STOP is obtained?

    b. We stop when all letters of the word STOP have been obtained?

**58. Alphonse and Beatrice are passengers on a ship. For the convenience of those interested in probability theory, the ship has on board an urn, some red balls, and some blue balls. Beatrice wins a bet from Alphonse. By the terms of the bet, the following is to be done: All the balls are to be placed in the urn. Alphonse is to draw balls, one at a time, with replacement, until a red ball is drawn. Alphonse is to pay Beatrice $10 for each ball drawn. However, before this process can be carried out, one of the red balls falls overboard. Beatrice proposes that they go ahead anyhow, but that, to compensate for the missing red ball, the drawing be without replacement. Is this fair?

5

More About Random Variables

The last chapter covered the basic facts about random variables. In this chapter we discuss a number of different topics related only in that all involve random variables. The first section covers some very important theoretical matters that are necessary for understanding the basic ideas of probability theory. The second section discusses the computation of expected values in certain circumstances; this material is used extensively in Chapters Eight and Nine. The optional third section is concerned with finding variances. The three sections may be read in any order.

5.1 The Law of Large Numbers

At the beginning of the book we expressed some rather vague ideas of what probability should mean. We broke off our discussion when it became clear we didn't know what we were talking about. We started anew in a more abstract way by discussing a set, the sample space, and a function that assigned numbers, that is, probabilities, to the subsets of that set. We made certain reasonable assumptions about the probability function. The remarkable thing is how well these few simple assumptions forced our theory to work out along the lines we envisaged. We are almost ready to make a precise statement about a coin falling heads half the time, but first we need a few more facts.

Let X be a random variable. We suppose that X has a variance and, consequently, a mean. We denote the mean of X by m and the standard deviation of X by σ; that is, we set $m = \mathbf{E}(X)$ and $\sigma^2 = \mathbf{Var}(X)$. We seek to study the probability that X takes a value "far" from m. We must first say what "far" means. In fact, an arbitrary standard of closeness will do. Let t be any positive number. We investigate how likely it is that X takes a value differing from m by t or more. In other words, we consider $P(|X - m| \geq t)$. The definition of variance is,

$$\sigma^2 = \mathbf{E}((X - m)^2).$$

Now using the definition of expected value, we have

$$\sigma^2 = \sum_{u \in \Omega} (X(u) - m)^2 \, P(u).$$

Let us consider those points in the sample space at which X takes a value "far" from m; more precisely, let A be the set of those u for which $|X(u) - m| \geq t$; we are studying $P(A)$. Let $B = \overline{A}$. Clearly then

$$\sigma^2 = \sum_{u \in A} (X(u) - m)^2 \, P(u) + \sum_{u \in B} (X(u) - m)^2 \, P(u).$$

Each term of the second sum is nonnegative; thus that sum itself is nonnegative. We have then

$$\sigma^2 \geq \sum_{u \in A} (X(u) - m)^2 \, P(u).$$

For $u \in A$, $|X(u) - m| \geq t$; thus $(X(u) - m)^2 \geq t^2$. It follows that

$$\sigma^2 \geq \sum_{u \in A} t^2 P(u) = t^2 \sum_{u \in A} P(u) = t^2 P(A).$$

Dividing by t^2, we have

$$\sigma^2 / t^2 \geq P(|X - m| \geq t).$$

This last inequality was first established by Irené-Jules Bienaymé (1795–1878). Nevertheless it is usually called the *Chebyshev Inequality*. We shall explain this name, and say more about Chebyshev, after we study the implications of the inequality.

It is easier to understand the Chebyshev Inequality if we write it in a slightly different form. Suppose we change the units in which we measure deviations from the mean and work in terms of the standard deviation. We may, for example, speak of deviating from the mean by two standard deviations, that is to say, by two times the standard deviation. The Chebyshev Inequality gives us information about the probability of deviating from the mean by at least c standard deviations. We rewrite the Chebyshev Inequality as

$$P\left(|X - m| \geq c\sigma\right) \leq 1/c^2,$$

for every $c > 0$; this form is obtained by simply setting $t = c\sigma$ in the original form of the Chebyshev Inequality.

Let us think about what the new form of the Chebyshev Inequality says. We do this by considering examples. First suppose $c = 2$. Then we have

$$P\left(|X - m| \geq 2\sigma\right) \leq 1/4.$$

In other words, no matter which random variable X we have, assuming it does have a variance, the probability that when we do the experiment X will miss its mean by as much as two standard deviations is no more than one-fourth. It is important to realize that for any particular X the probability just mentioned will "most likely" be way below one-fourth; one-fourth covers the worst possible case. With any additional information available, we could quite possibly make a much stronger statement about the probability. The extraordinary thing is that we can say anything at all that covers all cases. Now suppose $c = 10$. Then the inequality

$$P\left(|X - m| \geq 10\sigma\right) \leq 1/100$$

makes it clear that it is unlikely for *any* random variable to miss its mean by as much as ten standard deviations. The specific number $1/100$ could quite possibly be much reduced if we had additional information. Finally we give a silly example. Let $c = 1/2$. Then we have

$$P\left(|X - m| \geq \sigma/2\right) \leq 4.$$

While this statement may be uninformative, it is not incorrect—the probability involved is in fact less than four. Thus, depending on the value we give to c, the statement made by the Chebyshev Inequality ranges from the sublime to the ridiculous.

We next give a direct application of the Chebyshev Inequality to Bernoulli trials. Let the random variable X be the number of successes in n Bernoulli trials. Then $\mathbf{E}(X) = np$ and $\mathbf{Var}(X) = npq$. Let Y be the fraction of the trials in which success is obtained; in other words, let Y be the number of successes divided by the number of trials. Thus $Y = X/n$, and we have

$$\mathbf{E}(Y) = \mathbf{E}(X/n) = (1/n)\mathbf{E}(X) = (1/n)np = p,$$
$$\mathbf{Var}(Y) = \mathbf{Var}(X/n) = (1/n^2)\mathbf{Var}(X) = (1/n^2)npq = pq/n.$$

Let t be a positive number. Using the Chebyshev Inequality and the fact that $p < 1$ and $q < 1$, we have

$$P(|Y - p| \geq t) \leq \frac{pq/n}{t^2} \leq \frac{1}{nt^2}.$$

The significance of this conclusion will be explained shortly.

Now we return to the ideas of Chapter One. There we discussed repeating an experiment many times and counting how often a certain event occurs. For practical purposes, all the repetitions together amount to one grand experiment. This one experiment consists of a system of Bernoulli trials with a success corresponding to the occurrence of the event considered in the original experiment. We switch our language entirely to the terminology of Bernoulli trials. Then the idea with which we started is that, if the number of trials is large, the fraction of trials on which we get a success should be close to p, the probability of success on each individual trial.

Now, avoiding words like "large" and "should," we state and prove a definite theorem; we shall follow the proof with explanation and discussion.

Theorem *Let positive numbers s and t be given. Then there is a number N with the following property: Suppose we have a system of n Bernoulli trials with $n \geq N$. Let a be the probability that the fraction of trials on which a success is obtained will fail to be within t of p. Then $a < s$.*

Proof Let the random variable X be the number of successes; let $Y = X/n$. Then a is the probability that Y differs from p by at least t. In other words, $a = P(|Y - p| \geq t)$. As we saw a couple of paragraphs back,

$$P\left(|Y - p| \geq t\right) \leq \frac{1}{nt^2},$$

that is,

$$a \leq \frac{1}{nt^2}.$$

Let $N = 1/st^2$. Then, for $n \geq N$, we have

$$a \leq \frac{1}{nt^2} \leq \frac{1}{Nt^2} = \frac{st^2}{t^2} = s,$$

completing the proof. □

Let us try to understand the import of what we just did. The claim is that a certain probability is less than s. The smaller s is, the more impressive the claim. We also say Y is likely to be within t of p. Again, the smaller t is, the more impressive the statement. The idea is that we may set as high a standard of certainty as we like, barring absolute certainty. Then we may set as high a standard of closeness as we like, barring absolute equality. Then when we have a sufficiently large number of Bernoulli trials, the fraction of the trials on which we get a success will be that likely to be that close to p.

The last theorem is called *Bernoulli's Theorem*, after the same Jakob B. for whom Bernoulli trials are named; he was the first to prove it. Bernoulli's Theorem is basic in understanding what probability is all about. Bernoulli himself is reputed to have thought about it for 20 years. While the reader will most likely not spend that long, the ideas involved here are worth a great deal of thought. Bernoulli's Theorem was historically the first example of a certain type of theorem. A theorem of this type is often called a *law of large numbers*, and, accordingly, Bernoulli's Theorem is sometimes called *Bernoulli's Law of Large Numbers*. A more typical example of this type of theorem appears below as our next theorem. It will then be seen that this next theorem is in a way simply a straightforward generalization of Bernoulli's Theorem. But before going on, we pause to discuss Bernoulli himself and his theorem.

In understanding the history of probability, it is important to note that the abstract ideas of the Chebyshev Inequality were not developed until a long time after Bernoulli proved his theorem. Bernoulli's proof involved very complicated estimates of the probabilities of various numbers of successes. While these computations were far from easy, they do not constitute Bernoulli's real achievement. As a matter of fact, as we shall see in the next chapter, DeMoivre was able to reach a much more detailed conclusion only a few years after Bernoulli published his theorem. But the basic idea of a law of large numbers remains of great importance.

Probability theory forms only a small part of the mathematical work of Bernoulli; this part was included in his book, *Ars conjectandi*. However, even in that book, Bernoulli included much that is no longer regarded as belonging to probability theory. The history of the book is somewhat complicated and involves four Bernoullis. When Jakob Bernoulli died in 1705, his manuscript was inherited by his brother, Nikolaus (1662–1716). Nikolaus turned the manuscript over to his son, also Nikolaus (1687–1759), who prepared it for publication. By the time the book was finally published, the first edition of Montmort's book on probability had already appeared. We shall refer to the correspondence between Johann Bernoulli (1667–1748), another brother of Jakob, and Montmort in Chapter Eight. Thus it is far from clear which Bernoulli did what.

Jakob Bernoulli

(a/k/a James B. and Jacques B.) (Swiss, 1654–1705)

The Bernoulli family left Antwerp in 1583, going first to Frankfort and then to Basel, to escape persecution as Protestants by the Spanish. Having settled in Switzerland, a country of many languages, the family continued to use many languages; the result is that the first problem in sorting out the many Bernoullis is that each of them is often referred to under a number of translations of what is basically the same name. Even more confusing is that they, as many families do, tended to use the same first name for different members of the family. The following table is intended to help the reader keep track of the names. The table includes only those Bernoullis we shall have occasion to mention in this book. We omit the obvious variations in spelling, for example, Jacob and Jacobi for Jakob, that often appear in print.

First Generation

Nikolaus (1623–1708)

Second Generation
(Sons of Nikolaus (1623–1708))

Jakob = James = Jacques (1654–1705)
Nikolaus (1662–1716)
Johann = John = Jean (1667–1748)

Third Generation

Nikolaus (1687–1759), son of Nikolaus (1662–1716)
Daniel (1700–1782), son of Johann (1667–1748)

The three most eminent Bernoullis were Jakob, Johann, and Daniel. They were, successively in that order, professors of mathematics at Basel. Jakob and Johann are often known as the "Bernoulli brothers." [The third brother, Nikolaus (1662–1716), was a senator and a painter, rather than a mathematician.] Daniel made many discoveries in mathematical physics, and the Bernoulli Principle is named for him. Let us now turn to Jakob, whose work in probability is more important than that of the others. Jakob became a mathematician despite his father, who insisted that he study theology. Jakob was forced to teach himself mathematics, and even to hide his mathematics books from his father. As a young man, Jakob taught mathematics to a blind girl; later he wrote a book on how to teach mathematics to the blind. Also in his youth, he travelled extensively, meeting the leading mathematicians and physicists of the day. In 1687, he became a professor at Basel. Jakob Bernoulli suggested the use of the word "integral," as a term in calculus, to Leibniz.

Again we return to the basic ideas. Our discussion of $\mathbf{E}(X)$ started by suggesting that if our experiment were repeated many times, and if all went well, the average of the values assumed by X on each of the trials would be close to $\mathbf{E}(X)$. Now we're ready to prove a theorem along those lines. As before, many repetitions of an experiment are—for practical purposes, even though we have not given repetition any mathematical meaning—equivalent to a single more complicated experiment. We consider random variables $X_1, \ldots, X_n$ on the sample space for this composite experiment. We may, if we like, think of $X_1, \ldots, X_n$ as corresponding respectively to n repetitions of something. We mean independent repetitions, and therefore we require $X_1, \ldots, X_n$ to be independent. While we may choose to think of $X_1, \ldots, X_n$ as "essentially identical," all we need is for them to have the same mean and the same variance. Our claim is that if n is large enough, the average

$$\frac{X_1 + \cdots + X_n}{n}$$

will almost certainly be close to $\mathbf{E}(X)$. More precisely, we claim the following theorem holds:

Theorem   *Let numbers m, σ, s, and t be given. Suppose s and t are positive. Then there is a number N with the following property: Let $X_1, \ldots, X_n$ be independent random variables with $n \geq N$. Suppose $\mathbf{E}(X_i) = m$ and $\mathbf{Var}(X_i) = \sigma^2$ for all i. Then*

$$P\left(\left|\frac{X_1 + \cdots + X_n}{n} - m\right| \geq t\right) \leq s.$$

Proof   Let

$$Y = \frac{X_1 + \cdots + X_n}{n}.$$

Then we have

$$\mathbf{E}(Y) = \frac{1}{n}\mathbf{E}(X_1 + \cdots + X_n)$$

$$= \frac{1}{n}\left[\mathbf{E}(X_1) + \cdots + \mathbf{E}(X_n)\right]$$

$$= \frac{1}{n}(m + \cdots + m) \qquad (n \text{ terms})$$

$$= m,$$

and we have

$$\mathbf{Var}(Y) = \frac{1}{n^2}\mathbf{Var}(X_1 + \cdots + X_n)$$

$$= \frac{1}{n^2}\left[\mathbf{Var}(X_1) + \cdots + \mathbf{Var}(X_n)\right]$$

$$= \frac{1}{n^2}(\sigma^2 + \cdots + \sigma^2) \qquad (n \text{ terms})$$

$$= \frac{\sigma^2}{n}.$$

Applying the Chebyshev Inequality, we have

$$P\left(|Y - m| \geq t\right) \leq \frac{\sigma^2}{t^2 n}.$$

Clearly, if n is large enough, σ^2/t^2n will be less than s, completing the proof. □

The theorem just proved was first proved by Chebyshev. We shall call it the Law of Large Numbers, though many other theorems along roughly the same lines share that name. Bernoulli's Theorem is a special case of this Law of Large Numbers. To see that, we simply consider Bernoulli trials and let X_i be 1 or 0 according to whether the ith trial results in success or failure. Now consider working the other way round: obtaining the Law of Large Numbers as a corollary of Bernoulli's Theorem. Suppose the random variables $X_1, \ldots, X_n$ of the law all have the same distribution and assume only finitely many values. Then the Law of Large Numbers follows from Bernoulli's Theorem. The reasoning involved is informally described at the beginning of Chapter Four, in the discussion immediately preceding the definition of an expected value. Anyone with experience in writing mathematical proofs involving limits should be able to translate the informal discussion into a formal proof. Jakob Bernoulli himself likely regarded this Law of Large Numbers as such an obvious consequence of his theorem as to not be worth mentioning.

Now we consider the contribution of Chebyshev. We begin with a remark on the name Chebyshev itself. Many spellings of this name have appeared in print. The confusion arises, of course, because of the need to transliterate the Cyrillic. Watch out especially for such forms as Tchebycheff that begin with a T. Turning now to Chebyshev's work on probability, we begin by noting that his applications of the inequality that bears his name are sufficient justification for naming the inequality after him. Chebyshev was the first to prove the Law of Large Numbers in the form we just gave; he gave the same proof, using the Chebyshev Inequality, as we did above. In the first place, Chebyshev's proof is a most elegant replacement for Bernoulli's complicated computations. What is more important, not only is the Law of Large Numbers, as stated above, more general than the version given by Bernoulli's Theorem, but Chebyshev's approach opens up a whole field of study, extending well beyond the discrete probability to which we limit ourselves in this book.

P.L. Chebyshev

Pafnuty Lvovich Chebyshev (Russian, 1821–1894)

Chebyshev was one of nine children of a retired army officer. His brother Vladimir became a distinguished artillery general. When the family moved to Moscow in 1832, Pafnuty finished his secondary education at home before entering Moscow University. At the university, he started on a long list of distinguished mathematical accomplishments. When the time came to seek employment, he went to St. Petersburg. Like his brother, he concerned himself with ballistics, although Pafnuty's concern was theoretical; he made computations for use by the army. Pafnuty Chebyshev was also most interested in designing machinery. Not only did he investigate the theoretical principles, but he actually made some calculating machines. He continued to turn out mathematical papers until his death.

Let us discuss for a moment the practical significance of the Law of Large Numbers. Consider an insurance company. We recall how insurance works. The company agrees, in return for being paid a certain amount, the premium, in advance, to assume responsibility for paying whatever amount, the loss, that may be lost through a contingent event. We may regard the loss as a random variable. When the company issues a policy, the premium must be less than the maximum possible value of the loss; otherwise it would be absurd to buy the policy. Thus, on the one policy, there is a chance the company will pay out more than they take in. Of course, no insurance company issues just one policy. The company's aim is to make an overall profit on a large number of policies. The key words are "large number." According to the Law of Large Numbers, if the company sells a large enough number of policies, they can be as good as certain that the average actual amount paid out per policy is essentially the expected value, in our technical sense, of the loss. The premium can thus be adjusted to guarantee the company a profit.

A similar situation holds for gambling casinos. Naturally, the casino sets the rules so that the expected value of its gain on each bet is positive. Nevertheless each patron who does not bet too many times has a chance to come out ahead. The Law of Large Numbers assures the casino of an overall profit on the large number of bets that it makes.

Exercises

1. Explain why the exercises for this section differ so much from the exercises for the other sections of the book.

2. Given a number $a \geq 1$, describe a random variable X such that
$$P\left(|X - \mathbf{E}(X)| \geq a\sigma\right) = 1/a^2,$$
where $\sigma^2 = \mathbf{Var}(X)$. Show that if random variables X_1 and X_2 both have this property, $\mathbf{E}(X_1) = \mathbf{E}(X_2)$ and $\mathbf{Var}(X_1) = \mathbf{Var}(X_2)$, then $P(X_1 = t) = P(X_2 = t)$ for all t.

3. We could, but will not, use the ideas above and the Chebyshev Inequality to find an estimate of the minimum number of times a coin must be tossed to be 99.44% sure that it falls heads between 49.9% and 50.1% of the time. Explain why one would expect the Chebyshev Inequality to give a very substantial overestimate. (In the next chapter we shall learn how to give a reliable estimate.)

4. Prove: Let numbers m, σ, and s be given. Suppose s and m are positive. Then there is a number N with the following property: Let $X_1, \ldots, X_n$ be independent random variables with $n \geq N$. Suppose $\mathbf{E}(X_i) = m$ and $\mathbf{Var}(X_i) = \sigma^2$ for all i. Then $P(X_1 + \cdots + X_n < 0) \leq s$.

5. Show that if in the Law of Large Numbers we replace the assumption that all the X_i have the same variance σ^2 by the assumption they all have variance less than some number σ^2, the law remains valid.

6. Show that if in the Law of Large Numbers we omit the assumption that all the X have the same mean and replace the m in the conclusion by $[\mathbf{E}(X_1) + \cdots + \mathbf{E}(X_n)]/n$, the law remains valid.

7. Combine the ideas of the last two exercises to find a generalized version of the Law of Large Numbers.

8. Combine the ideas of exercises 4, 6, and perhaps 5 to find a statement about random variables with positive means.

5.2 Conditional Probability

As a preliminary to the next topic, we now work an easy exercise. Letters are chosen, one at a time, with replacement, from the word

DEFINITE

until we get a vowel. Let X be the number of times a letter is chosen. Then, since we have Bernoulli trials and seek one success, $\mathbf{E}(X) = 1/p = 2$. If instead of the word DEFINITE we use one of the words

ZERO, ONE, TWO, FIVE, SIX, SEVEN,

which one to be determined by chance in some manner unknown to us, then it is clearly impossible for us to find $\mathbf{E}(X)$.

Now we give an example illustrating the next topic. Two pennies and a nickel are tossed. The value in cents of the coins that fall heads is described by one of the words ZERO, ONE, TWO, FIVE, SIX, SEVEN. From that word we choose letters, one at a time, with replacement, until we get a vowel. Let X be the number of times a letter is chosen. It is almost obvious, on an informal basis, how to compute $E(X)$. The probabilities of the respective words being used are 1/8, 1/4, 1/8, 1/8, 1/4, 1/8. While we don't have a single value for X for each word, we do have an average value. For example, if the word is ZERO, it takes an average of two tries to get a vowel. Now it seems reasonable that, in computing $E(X)$, we need not distinguish between an average of two and a flat two. The cases of the other five words are similar. Thus we guess that

$$E(X) = (1/8)2 + (1/4)(3/2) + (1/8)3 + (1/8)2 + (1/4)3 + (1/8)(5/2).$$

We shall justify this equation shortly. But first we must decide precisely what the numbers 2, 3/2, 3, 2, 3, 5/2 really represent.

Just as we consider the conditional probability $P(B \mid A)$ of the event B on the assumption that the event A is known to happen, we can consider the conditional expectation $E(X \mid A)$. We define $E(X \mid A)$ by modifying the definition of $E(X)$ in the obvious way. By definition,

$$E(X) = \sum_{u \in \Omega} X(u)P(u).$$

We simply replace the probabilities $P(u)$ by the conditional probabilities $P(u \mid A)$ and define

$$E(X \mid A) = \sum_{u \in \Omega} X(u)P(u \mid A).$$

Of course, this definition makes sense only when $P(A) \neq 0$. Informally speaking, $E(X \mid A)$ may be described as follows: We do our experiment many times and each time note whether A happens. We record the value of X only on those occasions when A does happen. Then the average of *those* values of X should be close to $E(X \mid A)$. This intuitive description of $E(X \mid A)$ corresponds exactly to the more extensive discussion of $E(X)$ with which we started the last chapter. Likewise, the formal theory of conditional expectation is completely analogous to that given above for expectation. For example, the equation,

$$E(X \mid A) = \sum_{t} tP(X = t \mid A),$$

holds for exactly the same reasons as the corresponding formula for $E(X)$. We proceed at once to a theorem that is not a trivial modification of what we have already done. This theorem often provides a convenient way to find $\mathbf{E}(X)$.

Theorem *Let $A_1, \ldots, A_n$ be a partition of Ω. Suppose none of the A_i have probability zero. Then*

$$\mathbf{E}(X) = P(A_1)\mathbf{E}(X \mid A_1) + \cdots + P(A_n)\mathbf{E}(X \mid A_n).$$

Proof Let $X_1, \ldots, X_n$ be random variables defined as follows:

$$
\begin{aligned}
X_i(u) = X(u) \quad & \text{if } u \in A_i, \\
X_i(u) = 0 \quad & \text{if } u \notin A_i.
\end{aligned}
$$

Since each $u \in \Omega$ belongs to just one A_i, $X = X_1 + \cdots + X_n$. Hence $\mathbf{E}(X) = \mathbf{E}(X_1) + \cdots + \mathbf{E}(X_n)$. To complete the proof, we need merely show that $\mathbf{E}(X_i) = P(A_i)\mathbf{E}(X \mid A_i)$ for all i. By definition,

$$\mathbf{E}(X \mid A_i) = \sum_{u \in \Omega} X(u)P(u \mid A_i).$$

We need to evaluate $P(u \mid A_i)$ for each particular u; let $B = \{u\}$. Then we have

$$P(u \mid A_i) = P(B \mid A_i) = \frac{P(B \cap A_i)}{P(A_i)}.$$

If $u \in A_i$, then $B \cap A_i = \{u\}$; if $u \notin A_i$, then $B \cap A_i = \emptyset$. Thus

$$
\begin{aligned}
P(u \mid A_i) = P(u)/P(A_i) \quad & \text{if } u \in A_i, \\
P(u \mid A_i) = 0 \quad & \text{if } u \notin A_i.
\end{aligned}
$$

For $u \in A_i$, we have $X(u) = X_i(u)$, and thus

$$X(u)P(u \mid A_i) = X_i(u)P(u)/P(A_i).$$

For $u \notin A_i$, we have $X_i(u) = 0$, and thus again

$$X(u)P(u \mid A_i) = X_i(u)P(u)/P(A_i),$$

since both sides are zero. Therefore,

$$\mathbf{E}(X \mid A_i) = \sum_{u \in \Omega} X(u)P(u \mid A_i) = \sum_{u \in \Omega} X_i(u)P(u)/P(A_i).$$

Multiplying by $P(A_i)$, we have

$$P(A_i)\mathbf{E}(X \mid A_i) = \sum_{u \in \Omega} X_i(u)P(u) = \mathbf{E}(X_i),$$

completing the proof. □

We now work a problem using the formula just derived. We consider the three words

MINNEAPOLIS, MINNESOTA, MISSISSIPPI.

We are going to draw letters, one at a time, with replacement, from these words until we get an I. But we shall observe the following rules: The first drawing will be at random from all the letters, with each of the 31 letters having the same chance to be drawn. Subsequent drawings, if there are any, will be from the same word from which the first letter was obtained, with each letter in that word having the same chance to be drawn. We seek the expected number of drawings to be made.

Given our earlier work, it is easy to say how long it will take to get an I if we know which word we are using. For example, since MINNEAPOLIS has 11 letters including two Is, the probability of an I on any one draw is 2/11. In this case, it takes an average of 11/2 draws to get an I. Similarly, using MINNESOTA, it takes, on the average, 9 draws to get an I; and using MISSISSIPPI it takes 11/4 draws to get an I. The theorem tells us how to combine 11/2, 9, and 11/4 into an overall average. The probabilities the first draw is from MINNEAPOLIS, from MINNESOTA, and from MISSISSIPPI are 11/31, 9/31, and 11/31. Thus the answer to our question is

$$\frac{11}{31} \cdot \frac{11}{2} + \frac{9}{31} \cdot 9 + \frac{11}{31} \cdot \frac{11}{4} = \frac{687}{124}.$$

Exercises

9. Two dice are thrown. The total is noted, and that number of coins are tossed. What is the expected number of heads to be obtained?

10. Two coins are tossed. A die is thrown for each coin that falls heads. What is the expected number of spots shown on the dice?

11. A pair of dice is thrown repeatedly until the total obtained on the first throw is obtained again. What is the expected number of throws necessary?

12. A die is thrown and the number noted. Then the die is thrown repeatedly until a number at least as high as the number obtained on the first throw is thrown. Find the mean number of times the die is thrown including the first throw.

13. In the game of craps, a player throws a pair of dice. The game ends with the first throw if a 2, 3, 7, 11, or 12 is thrown. Otherwise, the player throws the dice repeatedly until either the number obtained on the first throw occurs again or a 7 is thrown. What is the expected number of throws to complete the game?

14. Two coins are tossed. If two heads are thrown, letters are chosen from the word TWOFOLD. If one head is thrown, letters are chosen from the word ONEROUS. If no heads are thrown, letters are chosen from the word NOMAD. Whichever word is used, we draw letters one at a time, with replacement, until a vowel is obtained. What is the expected number of letters then drawn?

15. Repeat the last exercise assuming the drawing to be without replacement.

16. The words

TATTLETALE, TATTOO, TATTING

contain a total of ten Ts. We choose one of those ten Ts at random. We continue to use the word containing that T and discard the other words. After replacing the T, we choose letters one at a time, with replacement, from the word in use until we get an A. On the average, how many times do we choose a letter, including the original choice of a T?

17. Repeat the last exercise assuming the drawing to be without replacement, except for the original T.

18. A die is thrown. Then cards are drawn from the deck, one at a time, with replacement, until the number of draws that have resulted in a spade is the number shown on the die. How many draws are made, on the average?

19. An urn contains two red balls and four green balls. Balls are drawn from the urn, one at a time, with replacement. The drawing continues until a red ball is drawn. If five or fewer of the draws result in green balls, no money changes hands.

a. If six or more draws result in green balls, Smith pays Jones one dollar each time a green ball is drawn, after the first five. Find the expected number of dollars Smith pays Jones.

b. If six or more draws result in green balls, Doe pays Roe one dollar each time a green ball is drawn, including a retroactive payment for the first five. Find the expected number of dollars Doe pays Roe.

20. The letters of the word

MISSISSIPPI

are arranged in random order. Let X be the number of Is that are immediately followed by an S. Find $E(X)$.

21. A hat contains eight one-dollar bills and two thousand-dollar bills. A coin is tossed. If it falls heads, Bill gets to draw, at random, two bills from the hat. If the coin falls tails, Bill draws only one bill, unless this first bill is a one-dollar bill; in the latter case, he gets to draw two additional bills. What is the average value of the bills Bill draws?

**22. One of the words RHUBARB and CABBAGE is chosen randomly. You choose letters at random from that word until you can be *sure* which word was chosen. What is the expected number of letters to be chosen if the choice is:

a. with replacement?

b. without replacement?

5.3 Computation of Variance

This section may be postponed indefinitely.
It is not needed in the rest of the book.

We have already made use of the technique of expressing a random variable as a sum of simpler random variables. We used this method to find the expected value of a random variable. However we hardly ever used the method to find a variance. The problem is, of course, that the variance of a sum need not be the sum of the variances. We shall see next that this difficulty can be easily overcome.

Before actually working a problem, let us get ready. Some terminology may help. We have seen that we often work with random variables of a certain kind. Specifically, let X be a random variable that assumes the value 1 when a certain event A happens and 0 when it does not happen. In this situation we call X the *indicator* of A. (Outside of probability theory, one would ordinarily say the same thing in different words by saying that X is the characteristic function of A.) If X is the indicator of A, then obviously $\mathbf{E}(X) = P(A)$, a fact we have often used. Suppose X and Y are the indicators of A and B. Since of course $1 \cdot 1 = 1$ and $0 \cdot 1 = 1 \cdot 0 = 0 \cdot 0 = 0$, we see that XY, like X and Y, assumes only the values 0 and 1. In detail, XY assumes the value 1 exactly when both X and Y assume the value 1. Thus, XY is the indicator of $A \cap B$, and hence $\mathbf{E}(XY) = P(A \cap B)$. Now we are ready for examples.

What are the mean and variance of the number of aces in a bridge hand (13 cards)? There are two alternate versions of the same basic method that we are about to use. We will do the problem twice for extra practice. Our first computation uses the events $A_1, \ldots, A_{13}$, where A_i is the event, "The ith card in the hand is an ace." Let $X_1, \ldots, X_{13}$ be the indicators of $A_1, \ldots, A_{13}$. Then $X_1 + \cdots + X_{13}$ is the number of aces in the hand; let $X = X_1 + \cdots + X_{13}$. As we have done before, we find

$$\mathbf{E}(X) = \mathbf{E}(X_1) + \cdots + \mathbf{E}(X_{13}) = 13\mathbf{E}(X_1) = 13 \cdot \frac{4}{52} = 1.$$

Now we try $\mathbf{Var}(X)$. Since we shall need $\mathbf{E}(X^2)$, we begin by studying X^2. Think for a moment about

$$X^2 = (X_1 + \cdots + X_{13})(X_1 + \cdots + X_{13}).$$

We see X^2 is the sum of all the products found by multiplying a term from the first factor by one from the second. Thus X^2 is the sum of the $X_i X_j$ for all i and j. It follows that $\mathbf{E}(X^2)$ is the sum of all the $\mathbf{E}(X_i X_j)$. As in finding $\mathbf{E}(X)$, we next use the fact that one X_i is very much like another. All $\mathbf{E}(X_i X_j)$ with $i \neq j$ are equal to each other; those with $i = j$ are also equal to each other. There are 13 ways to choose i and j with $i = j$ and $13 \cdot 12$ ways to choose i and j with $i \neq j$.

Thus we have

$$E(X^2) = 13E(X_1^2) + 13 \cdot 12E(X_1X_2),$$

where, essentially as we have done before, we use X_1 and X_1X_2 as examples since all X_i are "the same." Referring to the last paragraph and noting $A_1 \cap A_1 = A_1$, we have

$$E(X^2) = 13P(A_1) + 13 \cdot 12P(A_1 \cap A_2).$$

We already used the fact that $P(A_1) = 4/52 = 1/13$. We have

$$P(A_1 \cap A_2) = P(A_1)P(A_2 \mid A_1) = \frac{4}{52} \cdot \frac{3}{51} = \frac{1}{13} \cdot \frac{1}{17}.$$

Thus

$$E(X^2) = 13 \cdot \frac{1}{13} + 13 \cdot 12 \cdot \frac{1}{13} \cdot \frac{1}{17} = \frac{29}{17}.$$

Now we conclude that

$$\begin{aligned} \mathbf{Var}(X) &= E(X^2) - [E(X)]^2 \\ &= \frac{29}{17} - 1 \\ &= \frac{12}{17}. \end{aligned}$$

As we said before, we are going to further illustrate the method by working the problem again. We emphasize that it does not make much difference which problem we do, as long as we use the method under discussion. Now return to the aces in the bridge hand. Let each of A_s, A_h, A_d, A_c be the event of the hand containing the designated ace. Let X_s, X_h, X_d, X_c be the indicators of A_s, A_h, A_d, A_c. Then X $= X_s + X_h + X_d + X_c$ is the number of aces in the hand;

$$E(X) = E(X_s) + E(X_h) + E(X_d) + E(X_c) = 4E(X_s) = 4P(A_s).$$

Since 13 of the 52 cards are in the hand, we have $P(A_s) = 13/52 = 1/4$. Hence $E(X) = 4(1/4) = 1$. Now, reasoning as we did earlier, we have

$$E(X^2) = 4P(A_s) + 4 \cdot 3P(A_s \cap A_h).$$

The first 4 is the number of ways to choose one of the four aces, and the $4 \cdot 3$ is the number of ways to choose two different aces in order.

We just noted that $P(A_s) = 1/4$. We have

$$P(A_s \cap A_h) = P(A_s)P(A_h \mid A_s).$$

If the hand contains the ace of spades, it also contains 12 other cards chosen from the 51 other cards; thus the chances are 12 in 51 that the ace of hearts is also in the hand. Thus, $P(A_h \mid A_s) = 12/51 = 4/17$. It follows that $P(A_s \cap A_h) = (1/4)(4/17) = 1/17$. We conclude

$$\mathbf{E}(X^2) = 4 \cdot \frac{1}{4} + 4 \cdot 3 \cdot \frac{1}{17} = \frac{29}{17}.$$

As before, we find $\mathbf{Var}(X) = 29/17 - 1 = 12/17$.

Now we can find the variance of many of the random variables we considered in the last chapter. We give an example here. Suppose a committee with six members is to be formed by randomly selecting six of the twelve senators from the New England states. Let X be the number of states represented on the committee. In the last chapter we found $\mathbf{E}(X)$; now we find $\mathbf{Var}(X)$. A standard trick will help. Let Y be the number of states not represented on the committee. Then $X + Y = 6$, since there are six states in all. It is easy to see that $\mathbf{E}(X) = 6 - \mathbf{E}(Y)$ and $\mathbf{Var}(X) = \mathbf{Var}(Y)$. Let each of $A_1, \ldots, A_6$ be the event that a different one of the states is *not* represented on the committee; let $Y_1, \ldots, Y_6$ be the indicators of $A_1, \ldots, A_6$. Then $Y = Y_1 + \cdots + Y_6$ and we have, similarly to the examples already worked,

$$\mathbf{E}(Y) = 6P(A_1),$$
$$\mathbf{E}(Y^2) = 6P(A_1) + 6 \cdot 5P(A_1 \cap A_2).$$

There are several ways to compute the probabilities we need. For example,

$$P(A_1) = \frac{\binom{10}{6}}{\binom{12}{6}} = \frac{5}{22},$$

$$P(A_1 \cap A_2) = \frac{\binom{8}{6}}{\binom{12}{6}} = \frac{1}{33}.$$

Substituting, we have

$$\mathbf{E}(Y) = 6 \cdot \frac{5}{22} = \frac{15}{11},$$

$$\mathbf{E}(Y^2) = 6 \cdot \frac{5}{22} + 6 \cdot 5 \cdot \frac{1}{33} = \frac{25}{11},$$

$$\mathbf{Var}(Y) = \frac{25}{11} - \left(\frac{15}{11}\right)^2 = \frac{50}{121}.$$

Thus we conclude

$$\mathbf{E}(X) = 6 - \frac{15}{11} = \frac{51}{11},$$

$$\mathbf{Var}(X) = \frac{50}{121}.$$

We may as well record a portion of the technique under consideration in the form of a formula. Since the proof of the formula is just a simple application of the discussion above, we leave this proof as an exercise. Let $X_1, \ldots, X_n$ be the indicators of the events $A_1, \ldots, A_n$. We suppose there are numbers s and t such that

$$P(A_i) = s \qquad \text{for all } i,$$
$$P(A_i \cap A_j) = t \qquad \text{whenever } i \neq j.$$

Then

$$\mathbf{Var}(X_1 + \cdots + X_n) = ns + n(n-1)t - n^2 s^2.$$

We give another illustration of the use of this last formula by using it to settle a question that has been around for a while. In fact, it has been around since we gave a formula for the variance of the binomial distribution. That formula covered drawing balls, with replacement, from an urn containing balls of two different colors. We now find a formula that works when we draw the balls without replacement.

An urn contains T balls, each of them either red or green. Suppose R of the balls are red and G of the balls are green; thus, $T = R+G$. We suppose n balls are to be drawn from the urn at random, without replacement. Let X be the number of red balls to be drawn. We now find $\mathbf{Var}(X)$. As usual, let the event A_i be "the ith ball drawn is red." Preparing to use the formula,

$$\mathbf{Var}(X) = ns + n(n-1)t - n^2 s^2,$$

we note that

$$s = P(A_i) = \frac{R}{T},$$

$$t = P(A_i \cap A_j) = \frac{R}{T} \cdot \frac{R-1}{T-1} \qquad \text{whenever } i \neq j.$$

Substituting, we find

$$
\begin{aligned}
\mathbf{Var}(X) &= \frac{nR}{T} + \frac{nR(n-1)(R-1)}{T(T-1)} - \frac{n^2R^2}{T^2} \\
&= \frac{nRT(T-1) + nRT(n-1)(R-1) - n^2R^2(T-1)}{T^2(T-1)} \\
&= \frac{nR[T(T-1) + T(n-1)(R-1) - nR(T-1)]}{T^2(T-1)} \\
&= \frac{nR(T^2 - Tn - TR + nR)}{T^2(T-1)} \\
&= \frac{nR(T-n)(T-R)}{T^2(T-1)}.
\end{aligned}
$$

Recalling $T = R + G$, we have

$$\mathbf{Var}(X) = \frac{n(T-n)}{T-1} \frac{R}{T} \frac{G}{T}.$$

We now record our result and give a name to the distribution involved. Let positive integers n, R, and G be given; set $T = R + G$. Suppose the random variable X is such that

$$P(X = k) = \frac{\binom{R}{k}\binom{G}{n-k}}{\binom{T}{n}},$$

for those integers k such that $0 \leq k \leq R$ and $0 \leq n - k \leq G$; we suppose $P(X = k) = 0$ for all other k. Then we say that X has a *hypergeometric distribution*. It is clear from the last paragraph that

$$\mathbf{E}(X) = \frac{nR}{T},$$

$$\mathbf{Var}(X) = \frac{n(T-n)}{T-1} \frac{R}{T} \frac{G}{T}.$$

The formulas just stated may be applied to give still another solution to the problem of determining the mean and variance of the number of aces in a bridge hand. Here we are choosing 13 cards out of 52. Thus, $n = 13$ and $T = 52$. There are, in all, 4 aces and 48 other cards. Thus, $R = 4$ and $G = 48$. We have

$$\mathbf{E}(X) = \frac{13 \cdot 4}{52} = 1,$$
$$\mathbf{Var}(X) = \frac{13 \cdot 39}{51} \frac{4}{52} \frac{48}{52} = \frac{12}{17}.$$

Exercises

23. In Example 3 of the last section of Chapter Four, about putting letters in envelopes, we introduced a random variable X. Find **Var**(X).

24. Find the variance of the amount of money the contestant wins in Exercise 44 of Chapter Four.

25. Find the variance of the random variable in each of parts a–h of Exercise 46 of Chapter Four.

26. Find the variance of the random variable in Exercise 48 of Chapter Four.

27. Find the variance of the random variable in Exercise 49 of Chapter Four.

28. Find the variance of the random variable in part a of Exercise 50 of Chapter Four.

29. Find the variance of the random variable in Exercise 51 of Chapter Four.

30. Find the variance of the random variable in Exercise 52 of Chapter Four.

31. Find the variance of the random variable in Exercise 54 of Chapter Four.

32. Find the variance of the random variable in part b of Exercise 55 of Chapter Four.

33. Find the variance of the random variable in Exercise 56 of Chapter Four.

34. Find the variance of the random variable in each part of Exercise 57 of Chapter Four.

35. In Example 2 of the last section of Chapter Four, about Jane drawing balls, we introduced a random variable X. Find **Var(X).

6

CHAPTER

Approximating Probabilities

Sometimes an approximation is more valuable than an exact answer. For example, consider the following question: A certain association has 4000 members; 2000 of them are men and 2000 are women. Exactly half the members will be randomly chosen to each receive one ticket to a special event; all members have the same chance of getting a ticket. What is the probability that exactly 1000 men and exactly 1000 women get tickets? The answer to the question is quite obviously

$$\frac{\binom{2000}{1000}^2}{\binom{4000}{2000}}.$$

But how large is that? For example, is it more or less than .001? That's far from obvious. We can rewrite the answer as follows:

$$\frac{\left(\frac{2000!}{(1000!)^2}\right)^2}{\frac{4000!}{(2000!)^2}} = \frac{(2000!)^4}{(1000!)^4 4000!}.$$

135

As we shall soon learn how to find out, 4000! has 12,674 digits. Using a computer to put the last fraction in lowest terms, we find that the numerator and denominator have about 1000 digits each. That doesn't help us. Since there is no way to express the exact answer in a form we can comprehend, we should seek an approximation. In this chapter, we consider several ways of approximating numbers that appear in the theory of probability.

6.1   The Poisson Distribution

We shall be particularly concerned with a certain very important situation. This is the case of Bernoulli trials when the number of trials is large. To put our conclusions in firm mathematical language, we discuss limits as n "tends to infinity." Our basic starting point is the expression

$$\binom{n}{k} p^k q^{n-k}$$

for the probability of k successes in n Bernoulli trials. We shall let n tend to infinity. What about p, q, and k? q is determined by p, since $q = 1 - p$ always; thus we need only decide about p and k.

First, suppose we fix p and k and let n tend to infinity. As a matter of common sense, and as we can easily prove using the Chebyshev Inequality, the probability of k successes—even the probability of k or fewer successes—is very small when the number of trials is large; a large enough number of trials will almost certainly produce a large number of successes. We can avoid this difficulty by letting p get smaller as n gets larger.

The obvious way to make p depend on n is to hold the expected value np fixed. After doing some abstract mathematics, we shall consider applications. Let a positive number m and a nonnegative integer k be given. We shall evaluate

$$\lim_{n\to\infty} \binom{n}{k} p^k q^{n-k},$$

where $p = m/n$ and $q = 1 - p$ depend on n. Let

$$A_n = \binom{n}{k} p^k q^{n-k}.$$

Then we have

$$A_n = \frac{n!}{k!(n-k)!} \left(\frac{m}{n}\right)^k q^{n-k} = \frac{n(n-1)\cdots(n-k+1)}{k!} m^k (1/n)^k q^{n-k}.$$

The numerator of the fraction to the right of the last equal sign contains k factors; thus we may multiply each of these factors by $1/n$ and delete the expression $(1/n)^k$. Now we have

$$A_n = \frac{\frac{1}{n} \cdot 1 \cdot \frac{1}{n}(n-1)\frac{1}{n}(n-2)\cdots\frac{1}{n}(n-k+1)}{k!} m^k q^n q^{-k}$$

$$= \frac{\left(1 - \frac{1}{n}\right)\left(1 - \frac{2}{n}\right)\cdots\left(1 - \frac{k-1}{n}\right)}{k!} m^k \left(1 - \frac{m}{n}\right)^n q^{-k}.$$

Now let $n \to \infty$. Keeping in mind that k and m are fixed, we see that each of the $k - 1$ expressions

$$1 - \frac{1}{n}, 1 - \frac{2}{n}, \ldots, 1 - \frac{k}{n}$$

tends to 1 and that

$$q^{-k} = \left(1 - \frac{m}{n}\right)^{-k} \to 1^{-k} = 1.$$

It is a standard result of calculus that

$$\left(1 - \frac{m}{n}\right)^n \to e^{-m}.$$

[One way to show this is to note

$$\log\left(1 - \frac{m}{n}\right)^n = \frac{\log\left(1 - \frac{m}{n}\right)}{1/n}$$

and apply L'Hospital's Rule; the rule is named for G.F.A. de l'Hospital (1661–1704), but it was devised by Johann Bernoulli.] Using these results in the last expression for A_n, we have

$$\lim_{n \to \infty} A_n = \frac{m^k}{k!} e^{-m}.$$

Before illustrating the use of the formula just derived, we discuss its history. The formula is called the *Poisson Approximation* after S.D. Poisson. Poisson was a successor to Laplace in many respects. There are reports that Laplace treated Poisson like a son. Certainly Poisson carried on Laplace's work, also ranging over a wide variety of subjects. Contributing to probability theory formed only a small, although important, part of what Poisson accomplished. Returning to the Poisson approximation, we find that DeMoivre, whose work runs through this whole chapter, was the first to approximate the probability of no successes in a large number of Bernoulli trials. In 1837, Poisson generalized DeMoivre's conclusion to an arbitrary number of successes, obtaining what we are calling the Poisson Approximation. Poisson also generalized Bernoulli's Law of Large Numbers, and it was Poisson who invented the name, "Law of Large Numbers."

S.D. Poisson

Simeón-Denis Poisson (French, 1781–1840)

Poisson was the son of a Justice of the Peace. Early in the French revolution, his father died, leaving the family in poverty. Poisson was rescued by an uncle, who was a surgeon. Poisson attempted to study medicine, but his primary education was too weak. Then he turned to mathematics, with better success. In 1798, Poisson became a student at the Ecole Polytechnique in Paris. There he so impressed Lagrange that the council of the school unanimously excused Poisson from the final examinations for graduation (Joseph-Louis Lagrange, 1736–1813). This was probably all to the good because Poisson's clumsiness in drawing diagrams would likely have caused him to fail the examinations. Immediately after completing his studies there, Poisson secured a junior position at the Ecole Polytechnique. He was still there, but by then a dean, when he died in 1840.

To give an idea of how the Poisson approximation may be used, we consider a specific example. Suppose we are interested in how many large earthquakes there will be next year. Since, so far, earthquakes have proved unpredictable, we may regard them, for present purposes, as due to chance. To find out how common earthquakes are, we consult *The Cambridge Encyclopedia of Earth Sciences* (Cambridge University Press, 1981). In that book, we find the statement, "Great earthquakes with magnitudes exceeding 8 [on the Richter scale] occur about once every five to ten years." For simplicity, we'll settle for five years. In other words, the average number of "great" earthquakes per year is 1/5. How many opportunities for earthquakes are there worldwide in a year? In other words, what is n? We don't know, and we don't care. As long as n is large and p is small, we may use

$$\frac{m^k}{k!} e^{-m}$$

as an approximation to the probability of just k successes. This formula involves m and k; it does not mention n and p. In the case at hand, $m = .2$. The values given by the formula, for certain values of k, are as shown in the following table.

k	Probability of k great earthquakes in a year
0	.81873
1	.16374
2	.01637
3	.00109
4	.00005

There are many other situations in which the Poisson approximation is useful. Those that are described in probability books are often whimsical. A case in point: How many chips are there in a chocolate chip cookie? We know cookies are made in large batches. Many chocolate chips are thrown into the dough, and each chip has a small chance to wind up in the cookie we are going to study. Thus the Poisson approximation may be used to find the chances for various numbers of chips. A famous example used data on the number of Prussian cavalrymen killed by the kick of a horse in each of 16 corps in each of the years 1875–1894. The data conform to the Poisson theory remarkably well. More important examples are misprints and defective items generally. If we suddenly get an excess of defective items over the number suggested by the Poisson approximation, it might be important to find out why. The number of customers entering a store during each of successive short time periods could well conform to the Poisson approximation.

In the examples just given, we are concerned only with how many times something happens. The simplest sample space on which to study that would be the set of nonnegative integers. Suppose we choose a positive number m and use the formula

$$P(k) = \frac{m^k}{k!} e^{-m}$$

to *define* $P(k)$ for each of $k = 0, 1, 2, \ldots$. Then we may define $P(A)$, for each event A, by

$$P(A) = \sum_{k \in A} P(k).$$

We need to check $P(\Omega) = 1$. In other words, we must show

$$e^{-m} + me^{-m} + \frac{m^2}{2!} e^{-m} + \frac{m^3}{3!} e^{-m} + \cdots = 1.$$

Clearly the sum on the left is equal to

$$e^{-m} \left(1 + m + \frac{m^2}{2!} + \frac{m^3}{3!} + \cdots \right).$$

We recognize the series in parentheses, and we recall that its sum is e^m; now we have $e^{-m}e^m = 1$, as required. Since the points of the sample space are numbers, it makes sense to consider the random variable $X(k) = k$ for all $k \in \Omega$. For this random variable we have

$$P(X = k) = \frac{m^k}{k!} e^{-m},$$

for $k = 0, 1, 2, \ldots$.

Any random variable X for which, for some number $m > 0$,

$$P(X = k) = \frac{m^k}{k!} e^{-m},$$

for $k = 0, 1, 2, \ldots$, is said to have a Poisson distribution. Since the probabilities just given total 1, $P(X = k)$ is necessarily 0 when k is not a nonnegative integer. We now find $\mathbf{E}(X)$. [Since we obtained the Poisson distribution as a certain limit, it is easy to guess $\mathbf{E}(X)$. But here we seek a proof.] We have

$$\mathbf{E}(X) = 0 \cdot P(X = 0) + 1 \cdot P(X = 1) + 2 \cdot P(X = 2) + \cdots.$$

$$= me^{-m} + 2\frac{m^2}{2!}e^{-m} + 3\frac{m^3}{3!}e^{-m} + \cdots.$$

Since $n/n! = 1/(n-1)!$, we have

$$\mathbf{E}(X) = me^{-m} + m^2 e^{-m} + \frac{m^3}{2!}e^{-m} + \frac{m^4}{3!}e^{-m} + \cdots.$$

$$= me^{-m}\left(1 + m + \frac{m^2}{2!} + \frac{m^3}{3!} + \cdots\right)$$

$$= me^{-m}e^{m}$$

$$= m.$$

We shall show in the next chapter that $\mathbf{Var}(X)$ is also m.

Exercises

1. Suppose a book of 200 pages contains 100 misprints distributed among the pages at random. Find the probability that page 72 contains exactly two misprints.

2. A book of 500 pages contains 750 misprints.

   a. What is the probability that a certain page contains no misprints?

   b. At least two misprints?

3. Suppose that the number of telephone calls an operator receives from 9:00 to 9:05 A.M. follows a Poisson distribution with mean 3. Find the probability that the operator will receive:

   a. no calls in that interval tomorrow.

   b. three or more calls in that interval the day after tomorrow.

4. Suppose raisin bread "averages" six raisins per slice. What is the probability that a slice contains at least three raisins?

5. There are an average of 60 traffic accidents per month (30 days) at a certain very dangerous intersection. Assuming that the daily number of accidents follows a Poisson distribution, find the probability that:

   a. there will be no accidents next November 15th at this intersection.

   b. there will be three or more accidents next November 20th.

6. Find the number of chocolate chips a cookie should contain on the average if it is desired that the probability of a cookie containing at least one chocolate chip be .99.

7. A Shakespearean scholar has stated the printer of Shakespeare's sonnets made "more than 40" errors in 154 sonnets. Assuming for convenience that "more than 40" means 44, find the probability that a certain sonnet contains n errors for each of $n = 0, 1, 2, 3, 4, 5$.

8. On the basis of the remark in the last problem, assume that Shakespeare's sonnets were first printed with an average of $44/154 = 2/7$ errors per sonnet. What is the probability that at least one sonnet contains at least three errors?

9. Ignore, for simplicity, the possibility of being born on February 29th. Assume that a person is just as likely to be born on any one date as on any other date, excluding February 29th, of course. Find the probability that exactly two of a group of 365 people were born on August 16th.

10. We make the same assumptions as in the last problem. Suppose 253 people are picked at random. (We use 253 since $e^{-253/365} = 1/2$ almost exactly. Use 1/2 in your computations.)

    a. What is the probability that no one in the group was born on January 1st?

    b. That at least two people were?

    c. What is the expected number of different birthdays of the group?

    d. The expected number of days which are each the birthday of at least two of the people?

11. Let X have a Poisson distribution with mean m. Show that the most likely value for X to assume is $[m]$, with the following

exception: If m is an integer, X is just as likely to assume the value $m - 1$ as to assume the value m.

12. A batch of 100 chocolate chip cookies is made using 1290 chocolate chips. What is the most likely number of chips in a randomly chosen one of the cookies?

13. If the probability that a cookie contains at least one raisin is .995, what is the most likely number of raisins in that cookie?

14. A book of 500 pages contains 900 misprints. What is the most likely number of misprints to be found on a randomly chosen page?

6.2 Stirling's Formula

In the problem with which we started this chapter, the main difficulty is to find the approximate size of $n!$ when n is large. We shall now find out how to do that. Let us begin by deciding in detail what we are looking for. Consider an example. 11! is roughly 40 million. If we subtract 11! from 40 million, we get 83,200. That might ordinarily seem to be quite a large number, but 83,200 is small when compared to either 11! or 40 million. Suppose a_n is the approximation we shall decide on for $n!$. We don't care if $a_n - n!$ is large for large n, as long as $(a_n - n!)/n!$ is close to 0. But, since

$$\frac{a_n - n!}{n!} = \frac{a_n}{n!} - 1,$$

that means that we need $a_n/n!$ to be close to 1.

It is customary to use a certain value for the a_n of the last paragraph. Note that we cannot say more than that. Obviously, since we are looking for something that is only an approximation, there will be many possibilities. In deciding among them, in addition to accuracy, we must consider simplicity. While it is easy and instructive to start with $n!$ and actually derive our approximation, the ideas involved in doing that have nothing to do with probability theory, and therefore we leave the derivation for the Appendix to this chapter. Here we just announce the customary approximation.

The statement

$$n! \sim n^n e^{-n} \sqrt{2\pi n}$$

is usually called *Stirling's Formula*; the sign $\sim$ may be taken as meaning the expression to the right of the sign approximates the one to the left for large n. More formally, the statement means that

$$\lim_{n \to \infty} \frac{n^n e^{-n} \sqrt{2\pi n}}{n!} = 1.$$

It might be better to call this statement Stirling's approximation, rather than Stirling's Formula. (We shall say something about who Stirling was shortly.) Knowing the value of the limit above is often helpful in working with theory. But in a practical case, for example, when $n = 50$, we need to have some idea how accurate our approximation is. We shall see in the Appendix to this chapter that

$$.92 < \frac{n^n e^{-n} \sqrt{2\pi n}}{n!} < 1 \ldots$$

for all n, and

$$.99 < \frac{n^n e^{-n} \sqrt{2\pi n}}{n!} < 1$$

for $n \geq 9$.

The appearance of the quantity π in the last paragraph raises a number of questions. One of them, how does π get involved here at all, must be deferred to the Appendix as to details. But we can point out that π appears "all over the place" in mathematics. A more practical question is why we use π when all we get is an approximation anyhow. Using the approximation 2.5 for $\sqrt{2\pi}$, we could have noted that

$$.919 < \frac{2.5 n^n e^{-n} \sqrt{n}}{n!} < 1$$

for all n, and

$$.988 < \frac{2.5 n^n e^{-n} \sqrt{n}}{n!} < 1$$

for $n \geq 9$. But we agree with DeMoivre, about whom we shall say more in the last section of this chapter. DeMoivre, like us, wanted an approximation to $n!$ to use in probability theory. He obtained the formula above, except that his version included a constant B that he could evaluate only approximately. To quote him, "... feeling at the same time that what I had done answered my purpose tolerably well, I desisted from proceeding farther, till my worthy and learned Friend Mr. *James Stirling*, who had applied himself after me to that inquiry, found that the Quantity B did denote the Square-root of the Circumference of a Circle whose Radius is Unity." DeMoivre goes on to explain the reason for using π even though an approximation is good enough. "But altho' it be not necessary to know what relation the number B may have to the Circumference of the Circle, provided its value be obtained, either by pursuing Logarithmic Series before mentioned, or any other way; yet I own with pleasure that this discovery, besides that it saved trouble, has spread a singular Elegancy on the Solution."

James Stirling

(a/k/a Stirling the Venetian), (Scottish, 1692–1770)

Stirling was born of a distinguished family in Stirlingshire. He ran into problems because his family were Jacobites, that is, they supported the claim of James the Old Pretender to the throne. At Oxford, even though Stirling was acquitted of the charge of "cursing King George," he was forced to leave, without a degree, after several years of study. In 1715, Stirling went to Venice, where he taught mathematics. While in Venice he published mathematical papers in England by submitting them through Newton. In Venice, Stirling uncovered some of the trade secrets of the glassblowers. As a result, he fled Venice in fear for his life in 1725. Aided by Newton, he returned to England and secured a good position at a school in London. In 1735, he turned from mathematics to more applied matters, becoming the manager of the Scots Mining Company. In 1752, the City of Glasgow decided to spend 10 million pounds to make the city a seaport. The first expenditure from this fund was to pay for a silver teakettle to reward Stirling for surveying the River Clyde.

Now we are ready to give a reasonable answer to the question with which we started this chapter. We found that

$$\frac{(2000!)^4}{(1000!)^4 4000!}$$

was an exact, but awkward, answer. Let us apply Stirling's Formula:

$$\frac{\left(2000^{2000}e^{-2000}\sqrt{4000\pi}\right)^4}{\left(1000^{1000}e^{-1000}\sqrt{2000\pi}\right)^4 4000^{4000}e^{-4000}\sqrt{8000\pi}}$$

$$= \frac{2000^{8000}e^{-8000}(4000\pi)^2}{1000^{4000}e^{-4000}(2000\pi)^2 4000^{4000}e^{-4000}\sqrt{8000\pi}}$$

$$= \frac{2^{8000}1000^{8000}(4000\pi)^2}{1000^{4000}1000^{4000}4^{4000}(2000\pi)^2\sqrt{8000\pi}}.$$

Since $2^{8000} = (2^2)^{4000} = 4^{4000}$, the last quantity reduces to

$$\frac{(4000\pi)^2}{(2000\pi)^2\sqrt{8000\pi}} = \frac{2^2}{\sqrt{8000\pi}} = \frac{1}{\sqrt{500\pi}}.$$

With a calculator, we find that $1/\sqrt{500\pi}$ is approximately .025. Thus the probability we seek is about 1/40. The event in question is unlikely, but not really astounding.

Exercises

15. Find the number of digits in 100!. (Actually, find the number of digits in the Stirling's Formula approximation to 100!. Note: $e^{100} = 10^{43.43}$ approximately.)

16. Use Stirling's Formula to approximate:

    a. $\binom{2n}{n}$.

b. 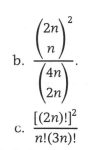 $\dfrac{\left(\dbinom{2n}{n}\right)^2}{\dbinom{4n}{2n}}.$

c. $\dfrac{[(2n)!]^2}{n!(3n)!}$

17. Let

$$x = \frac{2000 \cdot 1998 \cdot 1996 \cdot 1994 \cdots 2}{1999 \cdot 1997 \cdot 1995 \cdot 1993 \cdots 1}.$$

a. Show that

$$x = \frac{\left(2^{1000}1000!\right)^2}{2000!}.$$

b. Use Stirling's Formula to approximate x.

18. Use Stirling's Formula to approximate the probability that a coin that is tossed 2000 times falls heads exactly 1000 times.

19. In how many ways can 10 different objects be selected from among 30 (disregarding the order of selection)? Answer approximately using Stirling's Formula.

20. Suppose X has a Poisson distribution with $E(X) = 10$. Use Stirling's Formula to approximate $P(X = 10)$.

21. In a system of Bernoulli trials with n, p, and q as usual, suppose np is an integer. Use Stirling's Formula to show that the probability of exactly np successes is approximately

$$\frac{1}{\sqrt{2\pi npq}}.$$

6.3 The Normal Distribution

Often we want to consider Bernoulli trials where n is large and p is not small. Our goal here will be just to describe how to approximate the probabilities involved in this situation. We not only relegate the proof of the theorem we use to the Appendix to this chapter, but we also leave the formal statement of the theorem to be given there.

Let X be the number of successes in n Bernoulli trials. We already pointed out that the probability of any particular number of successes is very small. Thus we consider instead the probability the number of successes falls in a given range. In other words, we seek to approximate $P(a \leq X \leq b)$ under the conditions described above. While, as we said, no proof will be given here, it may help the reader to remember what to do if we give some indication of why the procedure used is plausible. It is not surprising that we want to see how far a and b are from $\mathbf{E}(X) = np$. It is also reasonable that we shall compare the deviations of a and b from $\mathbf{E}(X)$ to the standard deviation σ of X; we shall express $a - np$ and $b - np$ as so many times $\sigma = \sqrt{npq}$. In short, we should consider

$$c = \frac{a - np}{\sqrt{npq}} \quad \text{and} \quad d = \frac{b - np}{\sqrt{npq}}.$$

The approximation to $P(a \leq X \leq b)$ we give is completely determined by c and d. As long as the given data lead to the same c and the same d, the probability remains approximately the same. We claim $P(a \leq X \leq b)$ is approximately

$$\frac{1}{\sqrt{2\pi}} \int_c^d e^{-x^2/2} \, dx.$$

Unfortunately, we're not quite done. The indefinite integral of $f(x) = e^{-x^2/2}$ cannot be given explicitly in terms of the functions studied in calculus. That doesn't really matter; we would expect to evaluate the function with the aid of a table in any case. Most tables, including the one given just before the exercises, give the value of

$$F(t) = \frac{1}{\sqrt{2\pi}} \int_0^t e^{-x^2/2} \, dx$$

for various positive values of t. We have, regardless of the signs of c and d,

$$\frac{1}{\sqrt{2\pi}} \int_c^d e^{-x^2/2} \, dx = \frac{1}{\sqrt{2\pi}} \int_0^d e^{-x^2/2} \, dx - \frac{1}{\sqrt{2\pi}} \int_0^c e^{-x^2/2} \, dx$$

$$= F(d) - F(c).$$

If c is negative, we still have the slight problem of finding $F(t)$ for negative values of t. But, since $x^2 = (-x)^2$, for all t we have

$$\int_0^t e^{-x^2/2}\,dx = \int_{-t}^0 e^{-x^2/2}\,dx = -\int_0^{-t} e^{-x^2/2}\,dx.$$

It follows that $F(t) = -F(-t)$ for all t. Thus for negative t, we find $F(t) = -F(-t)$ by looking up $-t$ in the table. [But don't forget the minus sign in front of "$F(-t)$" in the last equation; actually picturing the area under the curve is a good precaution.] The table that is included here is a very short one, just long enough for our exercises. Of course, far more extensive tables are readily available. Most such tables have a title indicating that they are tables of the "Standard Normal Distribution." The word "distribution" is being used here in a more general sense than we have been using it. The standard normal distribution is a continuous distribution and thus falls outside of our topic of discrete probability. The approximation we have just introduced is often called the *Normal approximation*. The theorem that justifies the use of this approximation is the simplest example of a class of theorems called *central limit theorems*.

The method we just described was devised by Abraham DeMoivre. He describes it in the second edition of his book, *The Doctrine of Chances*, in language as informal as that we used. DeMoivre suggests power series and the Newton–Cotes formulas as ways to find approximate values for the definite integrals involved here; of course, tables of the values of these integrals had yet to be constructed. (indexNewton, IsaacNewton refers to the obvious person; Cotes refers to Roger Cotes, 1682–1716.) It seems appropriate to us to call the result simply DeMoivre's Theorem, as is often done, rather than to mention Laplace, as is also often done. In this respect, as well as others, DeMoivre had by 1738 made very substantial progress beyond the work of Montmort, about whom we shall say more in Chapter Eight, and Jakob Bernoulli. Without minimizing the theoretical importance of Bernoulli's Law of Large Numbers, we do point out that its use is confined to theory. It says certain probabilities are "small" when the number of trials is "large," but it does not state how small or how large. It gives no hint of how to say anything about probabilities that are not small. DeMoivre gives an approximation, sufficiently accurate for practical purposes, to the probabilities of different numbers of successes in a fixed number of Bernoulli trials. In particular, Bernoulli's Theorem is a corollary of DeMoivre's.

Before discussing the life of DeMoivre, we make a few more comments about his book. The first edition appeared in 1717 and the very significantly improved second edition in 1738, as we said. It is like those of Jakob Bernoulli and Montmort in that it begins with basic principles and goes on to work problems about games of chance. But DeMoivre, in part building on the work of his predecessors, goes well beyond them. The most important achievements are DeMoivre's Theorem, which was discussed in the last paragraph, and the use of generating functions to sum infinite series. The book includes the first explicitly worked out practical application of probability theory. (Of course, we do not count gambling as practical.) Specifically, building on the pioneering work of Edmund Halley (1656–1742) in devising a mortality table, DeMoivre includes the first table showing the present value of a life annuity; he also includes other actuarial tables. (As a matter of interest, we note that Halley is indeed the man who studied what we call Halley's comet.)

Abraham DeMoivre

(English, 1667–1754)

Abraham DeMoivre was born in France. His father, a surgeon, sent him away to school in 1673. Abraham was proud throughout his life that he was able to write a letter to his parents at that time. Like Jakob Bernoulli, he was forced to try to keep his study of mathematics secret. At one point, his teacher asked him what the "little rogue meant to do with all those cyphers." When the Edict of Nantes, which provided for the toleration of Protestants in France, was revoked in 1685, DeMoivre fled to England. It appears that he added the "de" to his name at this time. (We are following the usual practice, when writing in English, of treating "de" as an integral part of the name for citizens of English-speaking countries, but not for others. Thus we say "DeMoivre" and place the name alphabetically under D. However, for example, we say "Fermat" and place the name under F.) DeMoivre supported himself by travelling from house to house doing tutoring. He took with him pages from Newton's *Principia*, which he studied in his spare moments. He met and impressed Halley, secretary of the Royal Society, who helped him get his work published and introduced him to Newton. In 1697, DeMoivre became a fellow of the Royal Society. He never got a professorship, despite the efforts of Leibniz to get him one at Cambridge. Nevertheless, a case can be made for his being the leading English mathematician of his day; Alexander Pope included these lines in his "An Essay on Man":

> "Who made the spider parallels design,
> Sure as DeMoivre, without rule or line?"

DeMoivre followed Halley in the study of mortality tables, life expectancy, and annuities. Late in life, DeMoivre subsisted by answering questions along those lines at a coffee house. Thus he may be said to be the first professional actuary. Towards the end of his life, he added membership in the academy in Paris to his many similar honors. At that time, he was spending more and more time sleeping, by one account as much as 23 hours a day. Finally, at the age of 87, there came a time when he failed to wake up at all.

$$I = \frac{1}{\sqrt{2\pi}} \int_0^t e^{-x^2/2}\, dx$$

t	I	$1/2 - I$
0	0	$5.0 \cdot 10^{-1}$
0.1	.0398278	$4.6 \cdot 10^{-1}$
0.5	.1914625	$3.1 \cdot 10^{-1}$
0.6744896	.25	$2.5 \cdot 10^{-1}$
1	.3413447	$1.6 \cdot 10^{-1}$
1.5	.4331928	$6.7 \cdot 10^{-2}$
2	.4772499	$2.3 \cdot 10^{-2}$
2.5	.4937903	$6.2 \cdot 10^{-3}$
3	.4986501	$1.3 \cdot 10^{-3}$
4	.4999683	$3.2 \cdot 10^{-5}$
5	.4999997133	$2.9 \cdot 10^{-7}$
6	.4999999990	$9.9 \cdot 10^{-10}$
8		$6.2 \cdot 10^{-16}$
10		$7.6 \cdot 10^{-24}$
20		$2.8 \cdot 10^{-89}$

Exercises

22. 4500 dice are thrown. Find the probability of getting between 775 and 800 "sixes."

23. 720 dice are thrown. Find the probability of getting between 100 and 130 "sixes."

24. Find the probability that, when 1620 dice are thrown, between 255 and 300 of them fall "six."

25. Find the probability that, when 2880 dice are thrown, between 430 and 450 of them fall "six."

26. 10,000 coins are tossed.

    a. What is the probability that between 4950 and 5050 of them fall heads?

    b. Between 4850 and 5150?

    c. Between 4995 and 5005?

27. By expressing, approximately, the probabilities as definite integrals, determine which is more likely—that a coin that is tossed 2500 times falls heads between 1215 and 1290 times or that a die that is thrown 720 times turns up "six" between 106 and 132 times.

28. By expressing, approximately, the probabilities as definite integrals, determine which is more likely—that a coin that is tossed 400 times falls heads between 182 and 214 times or that a die that is thrown 180 times falls "six" between 24 and 39 times.

29. The probability that a randomly selected voter will answer "Yes" to a certain political question is 5/9. 720 voters are chosen at random, and each is asked the question. What is the probability that between 380 and 420 answer "Yes"? More than 360 answer "Yes"?

30. Suppose just two-thirds of all voters support a certain proposition. 800 voters are chosen at random. Find the probability that a majority of these 800 oppose the proposition.

31. An examination consists of 100 true-false questions. Find the probability that a student who answers all questions at random gets at least 65 right.

Appendix

We now prove Stirling's Formula. We begin by fixing our attention on some one integer $n \geq 3$ and trying to estimate $n!$. The idea behind our work is that

$$\log n! = \log n + \log(n-1) + \log(n-2) + \cdots + \log 1.$$

Towards studying the sum on the right, we consider the graph of $y = \log x$. A portion of that graph appears in each of the following three diagrams, the first of which applies for each integer $i \geq 3$ and the second for integers $i \geq 2$.

From the first of these diagrams we see, for each integer $i \geq 3$,

$$\int_{i-1}^{i} \log x \, dx \geq \frac{1}{2}[\log(i-1) + \log i].$$

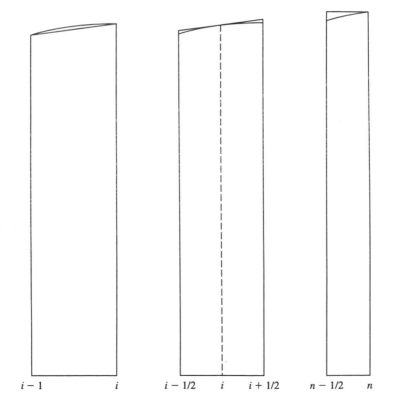

$i-1$ i $i-1/2$ i $i+1/2$ $n-1/2$ n

If we set, for brevity,

$$A_i = \int_{i-1}^{i} \log x \, dx - \frac{1}{2}[\log(i-1) + \log i]$$

for each $i = 3, 4, 5, \ldots$, then we have $A_i \geq 0$. We also set

$$A_2 = \int_{1}^{2} \log x \, dx - \frac{1}{2} \log 2.$$

Now let

$$B_2 = A_2$$
$$B_3 = A_2 + A_3$$
$$B_4 = A_2 + A_3 + A_4$$
$$\vdots$$

Clearly, $B_2 \leq B_3 \leq \ldots$. Let us compute B_n:

$$B_n = A_2 + \cdots + A_n$$
$$= \int_{1}^{2} \log x \, dx + \int_{2}^{3} \log x \, dx + \cdots + \int_{n-1}^{n} \log x \, dx - \frac{1}{2} \log 2$$
$$- \frac{1}{2}(\log 2 + \log 3) - \cdots - \frac{1}{2}[\log(n-1) + \log n]$$
$$= \int_{1}^{n} \log x \, dx - [\log 2 + \log 3 + \cdots + \log(n-1)] - \frac{1}{2} \log n.$$

Now consider the second diagram. From it we see,

$$\int_{i-1/2}^{i+1/2} \log x \, dx \leq \log i.$$

Thus, for each integer $n \geq 2$, we have

$$\int_{3/2}^{n-1/2} \log x \, dx = \int_{3/2}^{5/2} \log x \, dx + \cdots + \int_{n-3/2}^{n-1/2} \log x \, dx$$
$$\leq \log 2 + \log 3 + \cdots + \log(n-1).$$

From the third diagram,

$$\int_{n-1/2}^{n} \log x \, dx \leq \frac{1}{2} \log n.$$

Adding this in, and adding

$$\int_1^{3/2} \log x \, dx$$

to both sides, we have

$$\int_1^n \log x \, dx \le \log 2 + \cdots + \log(n-1) + \frac{1}{2}\log n + \int_1^{3/2} \log x \, dx;$$

$$\int_1^n \log x \, dx - [\log 2 + \cdots + \log(n-1)] - \frac{1}{2}\log n \le \int_1^{3/2} \log x \, dx.$$

But the left side is just B_n; thus,

$$B_n \le \int_1^{3/2} \log x \, dx.$$

for all n.

We have shown that $B_2 \le B_3 \le \ldots$, but that $B_n \le$ a certain number fixed for all n. Thus the sequence $B_2, B_3, \ldots$ must converge to some number. Call this number C; then we have, in symbols, $B_n \to C$.

We can actually evaluate the integral that appears in the expression for B_n. We have

$$\int_1^n \log x \, dx = (x\log x - x)\,|_1^n = n\log n - n - (1\log 1 - 1) = n\log n - n + 1.$$

Also we have

$$\log 2 + \log 3 + \cdots + \log(n-1) + \frac{1}{2}\log n = \log n! - \frac{1}{2}\log n.$$

Thus

$$B_n = n\log n - n + 1 - \log n! + \frac{1}{2}\log n.$$

Our goal was information about $n!$. We have

$$n! = e^{\log n!}$$
$$= e^{-B_n + n\log n - n + 1 + (1/2)\log n}$$
$$= e^{-B_n} e^{n\log n} e^{-n} e e^{\log \sqrt{n}}$$
$$= e^{-B_n} e n^n e^{-n} \sqrt{n}.$$

Hence, since $B_n \to C$,

$$n! / \left(n^n e^{-n} \sqrt{n} \right) \to e^{-C} e.$$

To summarize, we have shown that there is a number k such that

$$\frac{n!}{n^n e^{-n} \sqrt{n}} \to k.$$

Looking back at the reasoning above, we can get a pretty good estimate of the value of k. Clearly

$$C \leq \int_1^{3/2} \log x \, dx.$$

Since this integral is .1082, we have $k = e^{1-C} \leq 2.4395$. (In this paragraph, the numbers shown to four decimal places are, of course, approximations accurate to that number of places.) On the other hand, $B_{10} \leq C$, the number 10 being somewhat arbitrary. Since $B_{10} = .0727$, $k \leq 2.5276$. Thus, if we use 2.5 in place of k, the error is less than 3% — good enough for most practical purposes. In fact, it is clear that the error in using $2.5 n^n e^{-n} \sqrt{n}$ for $n!$ is less than 3% for all $n \leq 10$. We are now in the same position DeMoivre was in before he consulted Stirling; in other words, we are done for practical purposes, although it is more elegant to actually evaluate k. It will turn out the evaluation of k is a by-product of the next theorem. Therefore we proceed immediately to the proof of that theorem while postponing showing that $k = \sqrt{2\pi}$. We shall write $\sqrt{2\pi}$ in place of k during the proof, since we are really considering two separate problems; to stick to what we actually know, each $\sqrt{2\pi}$ in the next proof should be replaced by a k.

The most important theorem of this chapter was not even stated, let alone proved. We now give the statement and a proof.

Theorem *Let c and d be numbers with $c < d$. Let p and q be positive numbers with $p + q = 1$. For each positive integer n, let P_n be the probability that in n Bernoulli trials with probability p of success on each trial the number of successes is between*

$$np + c\sqrt{npq} \quad \text{and} \quad np + d\sqrt{npq}.$$

Then

$$\lim_{n \to \infty} P_n = \frac{1}{2\pi} \int_c^d e^{-x^2/2} \, dx.$$

Proof We are concerned with getting between $np + c\sqrt{npq}$ and $np + d\sqrt{npq}$ successes. When n is large enough, this range includes only numbers between 0 and n. More precisely, suppose

$$n \geq \frac{c^2 q}{p} \qquad \text{and} \qquad n \geq \frac{d^2 p}{q}.$$

From the first of these inequalities, we have $\sqrt{np} \geq |c|\sqrt{q} \geq -c\sqrt{q}$, and hence $np \geq -c\sqrt{npq}$. It follows that $np + c\sqrt{npq} \geq 0$. From $n \geq d^2 p/q$, we see in the same way that $nq \geq d\sqrt{npq}$. Thus we have $np + d\sqrt{npq} \leq np + nq = n$.

Since we are concerned only with a limit as $n \to \infty$, we may set a minimum size for n. From now on, we shall always assume that n satisfies the conditions of the last paragraph. We also assume that n is large enough so that there is at least one integer between $np + c\sqrt{npq}$ and $np + d\sqrt{npq}$.

Next we make a remark as to notation. The letter n will, of course, take different values during our discussion. Various other variables, k, t, R, S, m, σ, T, Δx, h, $k_1, \ldots, k_h$, $x_1, \ldots, x_h$, will be introduced and defined in terms of n. We shall not show the dependence of these variables on n in our notation. On the other hand, c, d, p, and q are constant and do not depend on n.

These preliminaries being completed, we now make a computation relating to a specific number k of successes for each number of trials. For each n, we choose an integer k such that

$$np + c\sqrt{npq} \leq k \leq np + d\sqrt{npq}.$$

We seek to study the probability of k successes in n trials, namely,

$$\binom{n}{k} p^k q^{n-k}.$$

We next introduce a very convenient variable t. For all n, let

$$t = \frac{k - np}{npq}.$$

Then we have

$$\frac{c}{\sqrt{npq}} \leq t \leq \frac{d}{\sqrt{npq}};$$

hence $t \to 0$ as $n \to \infty$. We put the Stirling's Formula approximation to

$$\binom{n}{k} p^k q^{n-k}$$

in terms of t. This approximation is RS, where

$$R = \frac{n^n e^{-n}}{k^k e^{-k}(n-k)^{n-k} e^{-(n-k)}} p^k q^{n-k}$$

and

$$S = \frac{\sqrt{2\pi n}}{\sqrt{2\pi k}\sqrt{2\pi(n-k)}}.$$

We treat R first. We have

$$R = \left(\frac{np}{k}\right)^k k \left(\frac{nq}{n-k}\right)^{n-k}$$

From $k = np + npqt$, we have

$$\frac{k}{np} = 1 + qt.$$

Also $n - k = n - np - npqt = nq - npqt$, and thus we have

$$\frac{n-k}{nq} = 1 - pt.$$

Thus

$$R = (1 + qt)^{-k}(1 - pt)^{-(n-k)}.$$

Integrating

$$\frac{1}{1+x} = 1 - x + x^2 - \cdots,$$

we obtain the well-known series for $\log(1 + x)$, namely,

$$\log(1 + x) = x - \frac{x^2}{2} + \frac{x^3}{3} - \cdots.$$

(That the last equation does not hold for all values of x is not important. We are only concerned with t close to zero.) Thus

$$\log R = -(np + npqt)\left(qt - \frac{(qt)^2}{2} + \frac{(qt)^3}{3} - \cdots\right)$$

$$- (nq - npqt)\left(-pt - \frac{(pt)^2}{2} - \frac{(pt)^3}{3} - \cdots\right)$$

$$= n\left(-pqt - pq^2t^2 + \frac{1}{2}pq^2t^2\right.$$

$$\left.+ \text{ terms involving powers of } t \text{ higher than the 2nd}\right)$$

$$+ n\left(pqt - p^2qt^2 + \frac{1}{2}p^2qt^2 + \text{terms involving powers, etc.}\right)$$

$$= -\frac{1}{2}n[pq(p + q)t^2 + \text{ terms, etc.}]$$

$$= -\frac{1}{2}n(pqt^2 + \text{ terms, etc.}).$$

We are concerned with the limit as $n \to \infty$, and hence $t \to 0$. R is thus approximated by

$$e^{-npqt^2/2} = e^{-(k-np)^2/2npq}.$$

Now we turn to S:

$$S = \frac{\sqrt{2\pi n}}{\sqrt{2\pi k}\sqrt{2\pi(n - k)}}$$

$$= \frac{1}{\sqrt{2\pi}}\sqrt{\frac{n}{k(n - k)}}$$

$$= \frac{1}{\sqrt{2\pi}}\sqrt{\frac{n}{(np + npqt)(nq - npqt)}}$$

$$= \frac{1}{\sqrt{2\pi n}}\sqrt{\frac{1}{(p + pqt)(q - pqt)}}.$$

For small t, we have approximately

$$S = \frac{1}{\sqrt{2\pi n}}\sqrt{\frac{1}{pq}} = \frac{1}{\sqrt{2\pi npq}}.$$

Putting everything together, we have just seen that the limit of the ratio of the probability of k successes in n trials, as described above, to

$$\frac{1}{\sqrt{2\pi npq}} e^{-(k-np)^2/2npq}$$

tends to 1 as $n \to \infty$.

At this point, some common abbreviations are helpful. Let $m = np$ and $\sigma^2 = npq$. Then the approximation in the last paragraph is

$$\frac{1}{\sigma\sqrt{2\pi}} e^{-(k-m)^2/(2\sigma^2)}.$$

The probability of between $m + c\sigma$ and $m + d\sigma$ successes in n trials is the sum of the probabilities of k successes for those integers k within this range. Denote these integers by $k_1, \ldots, k_h$ with $k_1 \leq \cdots \leq k_h$; note that we obtain entirely different sets of integers for different values of n. The sum of probabilities just mentioned is, according to the last paragraph, approximated by

$$\frac{1}{\sigma\sqrt{2\pi}} \sum_{i=1}^{h} e^{-(k_i-m)^2/(2\sigma^2)}.$$

Now turn to the integral

$$\int_{c}^{d} e^{-x^2/2}\, dx.$$

For each positive integer n, define Δx by $\Delta x = 1/\sigma$. We approximate the integral by dividing the interval from c to d into pieces, all but the last of length Δx. More formally, we approximate the integral by a *Riemann sum*. [These sums are named for Bernhard Riemann (1826–1866), who developed a formal theory of integration well over 100 years after DeMoivre, and many others, used such sums.] Let

$$x_i = \frac{k_1 - m}{\sigma},$$

for each $i = 1, \ldots, h$. Now $k_1, \ldots, k_h$ include in order all the integers between $m + c\sigma$ and $m + d\sigma$. Thus, $k_1 - 1 < m + c\sigma$, $k_h + 1 > m + d\sigma$, and $k_i = k_{i-1} + 1$ for all $i = 2, \ldots, h$. Using $k_1 - m < c\sigma + 1$, we have

$$x_1 = \frac{k_1 - m}{\sigma} < \frac{c\sigma + 1}{\sigma} = c + \frac{1}{\sigma} = c + \Delta x.$$

Likewise $k_h - m > d\sigma - 1$ yields $x_h > d - \Delta x$. Also

$$x_i - x_{i-1} = \frac{k_i - m}{\sigma} - \frac{k_{i-1} - m}{\sigma} = \frac{k_i - k_{k-1}}{\sigma} = \frac{1}{\sigma} = \Delta x.$$

Finally, note that $k_1 \geq m + c\sigma$ implies that $x_1 \geq c$; similarly $x_h \leq d$. Thus we have

$$
\begin{array}{lll}
c & \leq x_1 & \leq c + \Delta x, \\
c + \Delta x & \leq x_2 & \leq c + 2\Delta x, \\
c + 2\Delta x & \leq x_3 & \leq c + 3\Delta x, \\
& \vdots & \\
c + (h-2)\Delta x & \leq x_{h-1} & \leq c + (h-1)\Delta x. \\
c + (h-1)\Delta x & \leq x_h & \leq d.
\end{array}
$$

In short,

$$\Delta x \sum_{i=1}^{h} e^{-x_i^2/2} + [d - c - (h-1)\Delta x - \Delta x]e^{-x_h^2/2}$$

is a Riemann sum for

$$\int_c^d e^{-x^2/2}\,dx.$$

Note that we had a slight problem adjusting the last piece, but, since we have

$$0 \leq d - c - (h-1)\Delta x \leq x_h + \Delta x - x_{h-1} = 2\Delta x$$

the adjustment tends to zero as n tends to infinity. Now recall from the last paragraph that, in our current notation,

$$\frac{1}{\sqrt{2\pi}}\Delta x \sum_{i=1}^{h} e^{-x_i^2/2}$$

approximates the probability we seek. For large n, this approximate probability must be arbitrarily close both to the probability and to

$$\frac{1}{\sqrt{2\pi}}\int_c^d e^{-x^2/2}\,dx.$$

This completes the proof. $\qquad\qquad\qquad\qquad\qquad\qquad\qquad\square$

However we left a gap in our reasoning earlier in this Appendix. When we first mentioned $\sqrt{2\pi}$, we simply announced without explanation that a certain constant had the value $\sqrt{2\pi}$. To stick to what we really have established, every appearance of $\sqrt{2\pi}$ in this chapter so far should be replaced by an unknown constant; call it k. Now we evaluate k. By choosing t large enough, we can make

$$\frac{1}{k} \int_{-t}^{t} e^{-x^2/2} \, dx$$

as close as we like to

$$\frac{1}{k} \int_{-\infty}^{\infty} e^{-x^2/2} \, dx.$$

Consider a system of Bernoulli trials with usual notation. The Chebyshev Inequality tells us that the probability of between $np - t\sqrt{npq}$ and $np + t\sqrt{npq}$ successes is at least $1 - 1/t^2$, no matter what t and n we use. Thus, for t large enough, this probability is close to 1 for all n. Having chosen t very large, we then can use the last theorem to choose n so large that the probability just mentioned is as close as we please to

$$\frac{1}{k} \int_{-t}^{t} e^{-x^2/2} \, dx.$$

Putting this all together, we see we must have

$$\frac{1}{k} \int_{-\infty}^{\infty} e^{-x^2/2} \, dx = 1$$

to make all these arbitrarily close approximations possible. Thus

$$k = \int_{-\infty}^{\infty} e^{-x^2/2} \, dx.$$

Let us sketch the calculus necessary to evaluate the above improper integral. For every number y we have

$$ke^{-y^2/2} = \int_{-\infty}^{\infty} e^{-(x^2+y^2)/2} \, dx.$$

It follows that

$$k \int_{-\infty}^{\infty} e^{-y^2/2} \, dy = \int_{-\infty}^{\infty} \int_{-\infty}^{\infty} e^{-(x^2+y^2)/2} \, dx \, dy.$$

The integral on the left is equal to k. Thus changing to a double integral over the entire plane P, we have,

$$k^2 = \int \int_P e^{-(x^2+y^2)/2} \, dA.$$

Using polar coordinates, we have

$$k^2 = \int_0^\infty \int_0^{2\pi} r e^{-r^2/2} d\theta \, dr = -2\pi e^{-r^2/2} \mid_0^\infty = 2\pi.$$

Thus $k = \sqrt{2\pi}$, as we announced long ago.

7

CHAPTER

Generating Functions

This chapter may be postponed indefinitely.
It is not needed in the rest of the book.

In this short chapter, we introduce a certain method of treating those random variables that, like most of those we study, take only non-negative integers as values. This method is useful in more general circumstances, and it will be easier to understand if we first present it in its natural setting. What we want to introduce can properly be called a tool, a trick, a toy, or a technique. A more helpful word is code. By describing a sequence of numbers in what appears to be a roundabout way, we often make it easier to work with the sequence.

Let $a_0, a_1, a_2, \ldots$ be a sequence of numbers. Note that it is convenient to denote the first term by a_0, not a_1. By definition, the *generating function* of the sequence is the function f defined by

$$f(z) = a_0 + a_1 z + a_2 z^2 + a_3 z^3 + \cdots,$$

provided this series converges for at least some nonzero values of z. The use of the letter z, instead of some other letter, is immaterial. For any sequence, the series converges for $z = 0$. If the series also converges for other values of z, we get a function f that may be expanded in a Maclaurin Series by the techniques of calculus (Colin

Maclaurin, 1698–1746). (Some readers may be relieved to know that we're not going to do that; we just want to know that it can be done.) Thus given f, we can recover the sequence by taking the coefficients from the Maclaurin Series. To know the generating function is, in theory, to know the sequence. All the properties of the sequence are somehow encoded into its generating function. This use of a function to study a sequence was devised by DeMoivre and further elaborated by Laplace.

We shall be almost exclusively concerned with a special case. Let X be a random variable that assumes only nonnegative integers as values. We consider the sequence, $P(X = 0)$, $P(X = 1)$, $P(X = 2), \ldots$. Its generating function,

$$f(z) = P(X = 0) + P(X = 1)z + P(X = 2)z^2 + \cdots,$$

is called the *generating function* of X. (Sometimes the term "probability generating function" is used when other generating functions are also under discussion.) To be sure that the sequence in fact has a generating function, we must be sure that the series converges for a nonzero value of z. But, for $z = 1$, we have

$$f(1) = P(X = 0) + P(X = 1) + P(X = 2) + \cdots.$$

Since we know X assumes no value not included in 0, 1, 2, ..., we know that the series converges to 1. Thus we see that X does have a generating function and that $f(1) = 1$.

We continue to suppose f is the generating function of X. In other words, we still have

$$f(z) = P(X = 0) + P(X = 1)z + P(X = 2)z^2 + \cdots.$$

In the following discussion, let us just assume for the time being that all the series involved converge and that $\mathbf{E}(X)$ and $\mathbf{Var}(X)$ exist. Let $a_k = P(X = k)$; then

$$f(z) = a_0 z + a_1 z + a_2 z^2 + \cdots.$$

Differentiating, we have

$$f'(z) = a_1 + 2a_2 z + 3a_3 z^2 + \cdots,$$
$$f''(z) = 2a_2 + 2 \cdot 3a_3 z + 3 \cdot 4a_4 z^2 + \cdots.$$

Thus

$$f'(1) = a_1 + 2a_2 + 3a_3 + \cdots$$
$$= P(X = 1) + 2P(X = 2) + 3P(X = 3) + \cdots = \mathbf{E}(X).$$

We also have

$$f''(1) = 2a_2 + 2 \cdot 3a_3 + 3 \cdot 4a_4 + \cdots,$$
$$f'(1) + f''(1) = a_1 + (1+1)2a_2 + (1+2)3a_3 + (1+3)4a_4 + \cdots$$
$$= a_1 + 2^2 a_2 + 3^2 a_3 + 4^2 a_4 + \cdots$$
$$= P(X = 1) + 2^2 P(X = 2) + 3^2 P(X = 3) + \cdots$$
$$= \mathbf{E}(X^2).$$

Thus $\mathbf{Var}(X) = f''(1) + f'(1) - [f'(1)]^2$.

Now we discuss, without proving anything, a certain technical point. Let X and f be as above. Since the definitions of $\mathbf{E}(X)$ and $\mathbf{Var}(X)$ may involve infinite series, these quantities are not defined for every X. Correspondingly, $f'(1)$ and $f''(1)$ are not always defined. It can be shown that $\mathbf{E}(X)$ exists if and only if $f'(1)$ exists. Likewise $\mathbf{Var}(X)$ exists if and only if $f''(1)$ exists, and consequently $f'(1)$ also exists. In short, the formulas,

$$\mathbf{E}(X) = f'(1),$$
$$\mathbf{Var}(X) = f''(1) + f'(1) - [f'(1)]^2,$$

may be used to determine the existence of $\mathbf{E}(X)$ and $\mathbf{Var}(X)$ as well as their values.

We can use the method of generating functions to find the mean and variance of many of the random variables we have considered earlier. Suppose that we are considering the number of Bernoulli trials necessary to get a success. Let X be this number of trials; then X has a geometric distribution. We shall find f, the generating function of X, and use it to find the mean and variance. We have seen that the probability of the first success occurring on the kth trial is $q^{k-1}p$; that is, $P(X = k) = q^{k-1}p$ for $k = 1, 2, \ldots$. Thus we have

$$f(z) = pz + qpz^2 + q^2 pz^3 + q^3 pz^4 + \cdots.$$

This is a geometric series with first term pz and ratio qz. It follows

$$f(z) = \frac{pz}{1 - qz},$$

$$f'(z) = \frac{p(1 - qz) - pz(-q)}{(1 - qz)^2} = \frac{p}{(1 - qz)^2}$$

$$f''(z) = \frac{(-2)p}{(1 - qz)^3}(-q) = \frac{2pq}{(1 - qz)^3}.$$

Now we put in $z = 1$:

$$f'(1) = \frac{p}{(1 - q)^2} = \frac{p}{p^2} = \frac{1}{p},$$

$$f''(1) = \frac{2pq}{(1 - q)^3} = \frac{2pq}{p^3} = \frac{2q}{p^2}.$$

Thus we have

$$\mathbf{E}(X) = \frac{1}{p},$$

$$\mathbf{Var}(X) = \frac{2q}{p^2} + \frac{1}{p} - \frac{1}{p^2} = \frac{2q + p - 1}{p^2} = \frac{q}{p^2},$$

since $q + p = 1$.

Now we suppose X has a Poisson distribution. That means that there is a number $m > 0$ such that

$$P(X = k) = \frac{m^k}{k!}e^{-m} \qquad \text{for } k = 0, 1, 2, \ldots .$$

Thus the generating function of X is given by

$$f(z) = e^{-m} + me^{-m}z + \frac{m^2}{2!}e^{-m}z^2 + \cdots .$$

We have

$$f(z) = e^{-m}\left(1 + mz + \frac{(mz)^2}{2!} + \frac{(mz)^3}{3!} + \cdots\right) = e^{-m}e^{mz} = e^{mz-m}.$$

It follows

$$f'(z) = me^{mz-m},$$
$$f''(z) = m^2 e^{mz-m},$$
$$f'(1) = m,$$
$$f''(1) = m^2.$$

Thus

$$\mathbf{E}(X) = m,$$
$$\mathbf{Var}(X) = m^2 + m - m^2 = m.$$

Before continuing our computations of important generating functions, we establish a general fact that we shall need. When the generating functions of independent random variable X and Y are known, it is very easy to find the generating function of $X + Y$. We next derive the appropriate formula. Let X and Y be independent and have generating functions f and g. Then $X + Y$ obviously has a generating function, since the sum of nonnegative integers is a nonnegative integer. To find the generating function of $X + Y$, we need to find $P(X + Y = n)$ for each n. For the event, $X + Y = n$, to occur

$$
\begin{aligned}
&\text{either } (X = 0 \quad \text{and} \quad Y = n)\\
&\text{or } (X = 1 \quad \text{and} \quad Y = n - 1)\\
&\text{or } (X = 2 \quad \text{and} \quad Y = n - 2)\\
&\qquad\qquad\vdots\\
&\text{or } (X = n \quad \text{and} \quad Y = 0)
\end{aligned}
$$

must occur. Thus we have

$$P(X+Y = n) = P(X = 0 \quad \text{and} \quad Y = n) + \cdots + P(X = n \quad \text{and} \quad Y = 0).$$

Since X and Y are independent, it follows that

$$P(X + Y = n) = P(X = 0)P(Y = n) + \cdots + P(X = n)P(Y = 0).$$

Now we can write down the generating function h of $X + Y$:

$$
\begin{aligned}
h(z) = {}& P(X = 0)P(Y = 0)\\
&+ [P(X = 0)P(Y = 1) + P(X = 1)P(Y = 0)]z\\
&+ [P(X = 0)P(Y = 2) + P(X = 1)P(Y = 1)\\
&\quad + P(X = 2)P(Y = 0)]z^2\\
&+ \cdots.
\end{aligned}
$$

Similarly for X and Y we have

$$f(z) = P(X = 0) + P(X = 1)z + P(X = 2)z^2 + \cdots,$$
$$g(z) = P(Y = 0) + P(Y = 1)z + P(Y = 2)z^2 + \cdots.$$

It can be shown that the obvious way to multiply these last two series gives the correct result, namely, the series given for $h(z)$ above. Thus the generating function $X + Y$ is fg, the product of the generating function of X and Y.

Now we consider the generating function for X, the number of successes in n Bernoulli trials. We first consider the special case where $n = 1$; that is, where X has a Bernoulli distribution. Thus we seek the number of successes in one trial. Then $P(X = 1) = p$ and $P(X = 0) = q$. It follows that the generating function of X is

$$g(z) = q + pz.$$

Now consider an arbitrary n. Then X has a binomial distribution. As usual, we set

$$X_i = \begin{cases} 1 & \text{if the } i\text{th trial results in success,} \\ 0 & \text{otherwise.} \end{cases}$$

We have $X = X_1 + X_2 + \cdots + X_n$ and $X_1, \ldots, X_n$ are independent. Each X_i has generating function $g(z) = q + pz$; thus X has generating function

$$f(z) = (q + pz)^n.$$

Now we turn to the random variable Y, the number of Bernoulli trials necessary to get r successes. In other words, suppose Y has a Pascal distribution. We already know the generating function is

$$g(z) = \frac{pz}{1 - qz}$$

when $r = 1$. Reasoning again as we just did, we see that Y has the generating function

$$f(x) = \left(\frac{pz}{1 - qz}\right)^r.$$

Our conclusions are summarized in the table, "Table of Important Distributions," at the end of the book. This table is the one we started drawing up in Chapter Four. Note that the mean and variance, shown in the table, for a Pascal distribution can now be obtained either from the generating function or by expressing the random variable as a sum, as in the last paragraph.

We next make an important point by working an example. In probability theory, as elsewhere, generating functions other than those of random variables can be very useful. Suppose someone offers you the following even-money bet: Two decks of cards are to be shuffled separately. Then the decks are placed next to each other on a table. The two top cards are compared; then the second from the top; then the third; and so on through all the cards. Your opponent wins only if an exact match, both as to face value and as to suit, is found at some place in the two decks. Otherwise you win. Should you bet?

Let us change the question into another form and generalize it. We have n letters, each addressed to someone by name. We also have an addressed envelope for each of the letters. If, instead of putting each letter in its own envelope, the letters are placed one in each envelope at random; what is the probability that just k letters are placed in the correct envelopes? In case $k = 0$, we are finding the probability of getting all the letters in the wrong envelopes. Thus the case where $n = 52$ and $k = 0$ is the card problem of the last paragraph. We shall denote the probability, for each n, of getting all n letters wrong by a_n. (Before reading further, guess how a_n varies with n. About how big is a_n for $n = 5, 10, 50, 100, 500$, etc.?) We, somewhat arbitrarily, set $a_0 = 1$. Clearly $a_1 = 0$; with only one letter and one envelope, there is no way to go wrong. However it is easier to write a_1 in some of our computations than to justify omitting a term. We first express the probability of getting k letters right with n envelopes in terms of the numbers $a_0, a_1, \ldots$; then we determine these numbers.

We begin by considering a fixed number n of letters and envelopes. a_n denotes the probability all the letters are put in the wrong envelopes. How about just one right? The probability that the first letter—we assume the letters are numbered from 1 to n in some arbitrary order—is in the right envelope is $1/n$. Assuming that the first letter is in the right envelope, the remaining $n - 1$ letters are distributed among the corresponding $n - 1$ envelopes. Thus, still assuming the first letter is right, the probability all the other $n-1$ letters are wrong is a_{n-1}. Combining these facts, we see that the probability that the first letter is the only one right is $(1/n)a_{n-1}$. The probability the second letter, or any other preselected letter, is the only one right is likewise $(1/n)a_{n-1}$. It follows that the probability that just one letter is correct, which one not being specified in advance, is $n(1/n)a_{n-1} = a_{n-1}$. Now we make the corresponding computation for two letters right. The probability the first letter is right is $1/n$. The probability the second letter is right, given that the first one is, is $1/(n-1)$. Given that the first two are right, the probability the others are all wrong is a_{n-2}. Thus the probability that the first two letters are the only ones that are right is $(1/n)[1/(n-1)]a_{n-2}$. The same would apply to any other two preselected letters. There are

$$\binom{n}{2} = \frac{n(n-1)}{2}$$

ways to select which two letters are to be right. Thus the probability that exactly two letters are in the correct envelopes is

$$\frac{n(n-1)}{2} \cdot \frac{1}{n} \cdot \frac{1}{n-1}a_{n-2} = \frac{a_{n-2}}{2}.$$

The analogous computation for three letters is, of course,

$$\binom{n}{3}\frac{1}{n} \cdot \frac{1}{n-1} \cdot \frac{1}{n-2}a_{n-3} = \frac{a_{n-3}}{3!}.$$

In general, the probability that just k letters are correct is $a_{n-k}/k!$ for $k = 0, 1, \ldots, n$. Thus, as soon as we find the values of $a_0, a_1, a_2, \ldots$, we shall have our whole problem solved.

The last paragraph not only reduces our problem to that of evaluating the a_i, but it also gives us a means of completing that evaluation. With n letters, the number of letters in the correct envelopes is some number from 0 to n. Thus the corresponding probabilities total 1. Thus we have

$$1 = a_n + a_{n-1} + a_{n-2}/2! + a_{n-3}/3! + \cdots + a_0/n!.$$

This is clearly of the form

$$a_n b_0 + a_{n-1}b_1 + a_{n-2}b_2 + \cdots + a_0 b_n,$$

which appeared above when we discussed the product of two generating functions, with $b_k = 1/k!$. Let f be the generating function of $a_0, a_1, a_2, \ldots$ and g be that of $1, 1, 1/2!, 1/3!, \ldots$. Then, from the total of the probabilities above being 1, it follows that fg is the generating function of 1, 1, 1, ... In other words,

$$f(z)g(z) = 1 + z + z^2 + z^3 + \cdots = 1/(1 - z).$$

Since

$$g(z) = 1 + z + \frac{z^2}{2!} + \frac{z^3}{3!} + \cdots = e^z,$$

we have $f(z)e^z = 1/(1 - z)$. We conclude that

$$f(z) = \frac{e^{-z}}{1 - z}.$$

It remains to actually evaluate $a_0, a_1, \ldots$ from their generating function f. We have

$$f(z) - zf(z) = e^{-z}.$$

The left side of the last equation is

$$(a_0 + a_1 z + a_2 z^2 + \cdots) - (a_0 z + a_1 z^2 + a_2 z^3 + \cdots)$$
$$= a_0 + (a_1 - a_0)z + (a_2 - a_1)z^2 + \cdots;$$

the right side is

$$e^{-z} = 1 - z + \frac{z^2}{2!} - \frac{z^3}{3!} + \cdots.$$

Thus we have

$$a_0 = 1,$$
$$a_1 - a_0 = 1,$$
$$a_2 - a_1 = 1/2!,$$
$$a_3 - a_2 = -1/3!,$$
$$a_4 - a_3 = 1/4!,$$
$$\vdots$$

We can read off

$$a_0 = 1,$$
$$a_1 = a_0 + (a_1 - a_0) = 0,$$
$$a_2 = a_1 + (a_2 - a_1) = 1/2!,$$
$$a_3 = a_2 + (a_3 - a_2) = 1/2! - 1/3!,$$
$$a_4 = a_3 + (a_4 - a_3) = 1/2! - 1/3! + 1/4!,$$
$$\vdots$$

We conclude

$$a_n = 1/2! - 1/3! + 1/4! - \cdots + (-1)^n/n!.$$

We can thus use

$$1 - 1 + 1/2! - 1/3! + 1/4! - \cdots = e^{-1}$$

to approximate a_n. The remarkable thing is how little a_n varies with n. The relative error in using $1/e$ in place of a_n is less than 2% for all $n \geq 4$. Assuming we have at least four letters, the probability of getting them all in the wrong envelopes is .37, to two decimal places, regardless of how many letters we have.

The problem just solved was first treated by Pierre Montmort. We shall have more to say about him at the beginning of the next chapter. Montmort used a different method from ours to find the probability of no matches. The problem is often referred to by the French name of *rencontre*.

Given n objects in a row, we may consider the number of rearrangements of the objects, still in a row, that leave no object in its original place. This number D_n is called the number of *derangements* of the objects. Clearly we have $D_n = n!a_n$. Using the properties of alternating series studied in calculus, we can see that D_n differs from $n!/e$ by less than $1/2$. In short, for all n, D_n is the integer closest to $n!/e$.

We note in passing an application of generating functions that is not directly related to probability theory. Suppose we need to find the sum of the series

$$\frac{1}{3} + \frac{2}{9} + \frac{3}{27} + \frac{4}{81} + \cdots.$$

Were it not for the numerators, 2, 3, 4,..., the problem would be easy. But since those numbers are there, we replace the problem by an apparently still harder problem. Evaluate

$$f(z) = \frac{1}{3} + \frac{2}{9}z + \frac{3}{27}z^2 + \frac{4}{81}z^3 + \cdots;$$

the special case $z = 1$ is the original problem. If we integrate each term and set

$$g(z) = \frac{1}{3}z + \frac{1}{9}z^2 + \frac{1}{27}z^3 + \frac{1}{81}z^4 + \cdots;$$

then $g'(z) = f(z)$. Since the series for $g(z)$ is a geometric series, we have

$$g(z) = \frac{(1/3)z}{1 - (1/3)z} = \frac{z}{3 - z}.$$

It follows

$$f(z) = g'(z) = \frac{(3 - z) - z(-1)}{(3 - z)^2} = \frac{3}{(3 - z)^2}.$$

Thus we have

$$\frac{1}{3} + \frac{2}{9} + \frac{3}{27} + \frac{4}{81} + \cdots = f(1) = \frac{3}{4},$$

answering the original question.

Let us try a somewhat different question. Find the sum of

$$\frac{1}{3 \cdot 4} + \frac{1}{4 \cdot 16} + \frac{1}{5 \cdot 64} + \frac{1}{6 \cdot 256} + \cdots.$$

Again there are some integers, 3, 4, 5, 6, ..., that are in our way. To remove these integers from the denominators, we shall have to differentiate, instead of integrating as we did in the last problem. We put in whatever powers of z will do the job of removing 3, 4, 5, 6, ... Then we have

$$f(z) = \frac{1}{3 \cdot 4} z^3 + \frac{1}{4 \cdot 16} z^4 + \frac{1}{5 \cdot 64} z^5 + \cdots;$$

again we need $f(1)$. Now we have

$$f'(z) = \frac{1}{4} z^2 + \frac{1}{16} z^3 + \frac{1}{64} z^4 + \cdots = \frac{(1/4)z^2}{1 - (1/4)z} = \frac{z^2}{4 - z}.$$

Integrating, to get back to $f(z)$, is a bit of a chore. We have

$$\frac{z^2}{4 - z} = -z - 4 + \frac{16}{4 - z}.$$

Thus

$$f(z) = -z^2/2 - 4z - 16\log(4 - z) + C.$$

(Don't forget the constant of integration!) To find C, note that obviously from its definition $f(0) = 0$. Thus

$$0 = -16\log 4 + C;$$

hence, $C = 16 \log 4$. We have then

$$f(z) = -z^2/2 - 4z - 16\log(4-z) + 16\log 4.$$

It follows

$$f(1) = -1/2 - 4 + 16\log(4/3) = 16\log(4/3) - 9/2,$$

and that is the answer to the problem we are solving.

We work another example, this one from probability theory. Let X be the number of Bernoulli trials necessary to get one success. We know $\mathbf{E}(X) = 1/p$. We now consider $\mathbf{E}(1/X)$, the average ratio of the number of successes, 1, to the number of trials, X. We have

$$\mathbf{E}(1/X) = P(X=1) + \frac{1}{2}P(X=2) + \frac{1}{3}P(X=3) + \cdots$$

$$= p + \frac{1}{2}qp + \frac{1}{3}q^2p + \cdots.$$

We sum this series by using generating functions. By analogy with the last example, we set

$$f(z) = pz + \frac{1}{2}qpz^2 + \frac{1}{3}q^2pz^3 + \cdots.$$

Then we have

$$f'(z) = p + qpz + q^2pz^2 + \cdots = \frac{p}{1-qz}.$$

$$f(z) = -p\log(1-qz)/q + C.$$

Since $f(0) = 0$ from the definition of $f(z)$, we find $C = 0$. Thus we have

$$\mathbf{E}(1/X) = f(1) = -(p/q)\log(1-q) = -(p/q)\log p.$$

The difference between $\mathbf{E}(1/X)$ and $1/\mathbf{E}(X)$ is not surprising in that an average of reciprocals is usually not the reciprocal of the average.

Exercises

1. Find the generating function of each of the following sequences:

   a. 3, 6, 12, 24, ...;

   b. 1, 3, 9, 27, ...;

c. $1, -3, 9, -27, 81, \ldots$;

d. $3, 9, 27, 81, \ldots$;

e. $0, 1, 3, 9, 27, \ldots$;

f. $1/4, 1/16, 1/64, 1/256, \ldots$;

g. $1, 1/3, 1/9, 1/27, \ldots$;

h. $2, 2, 2, 2, \ldots$;

i. $1, -1/2, 1/4, -1/8, \ldots$.

2. Suppose a coin is tossed repeatedly until it falls heads. Let X be the number of tosses necessary. Find the generating function of X.

3. A coin is tossed repeatedly until either tails appears or heads has appeared four times, whichever comes first. Find the generating function of the number of tosses necessary.

4. Exactly one of six similar keys is known to open a certain door. They are tried one after another, and X is the number of tries necessary to open the door. Find the generating function of X:

a. if a key that doesn't work is discarded;

b. if a key that doesn't work is mixed in with the others and may be tried again.

**5. Let f be the generating function of the random variable X. In other words, if we define $a_n = P(X = n)$ for each $n = 0, 1, 2, \ldots$, then $f(z) = a_0 + a_1 z + a_2 z^2 + \cdots$. For each of the following sequences $b_0, b_1, b_2, \ldots$, write out the first four nonzero terms and find the generating function g of this sequence.

a. $b_n = 1$ for all n.

b. $b_n = P(X \neq n) = 1 - P(X = n)$.

c. $b_n = P(X = n + 1)$.

d. $b_n = P(X = n - 1)$.

e. $b_n = P(X \leq n) = P(X < n + 1)$.

f. $b_n = P(X > n) = P(X \geq n + 1) = 1 - P(X \leq n)$.

g. $b_n = P(X \leq n - 1) = P(X < n)$.

h. $b_n = P(X > n - 1) = P(X \geq n)$.

6. A random variable X has generating function

$$f(z) = 1/4 + (1/3)z + (1/6)z^2 + (1/6)z^3 + (1/12)z^4.$$

Find:

a. $P(X = 2)$;

b. $P(X = 0)$;

c. $P(0 < X < 3)$;

d. $P(X > 1)$.

7. Suppose X and Y are independent random variables and each has a Poisson distribution. Show that $X + Y$ has a Poisson distribution by finding the generating function of $X + Y$. [Note: $E(X)$ need not equal $E(Y)$.]

8. Let X be the number of spots obtained when a die is thrown. Let Y be the number of heads obtained when two coins are tossed. Find the generating function of:

a. X;

b. Y;

c. $X + Y$;

d. $2X$.

9. A random variable X has the generating function

$$f(z) = \frac{1}{(2 - z)^2}.$$

Find:

a. $P(X = 0)$;

b. $E(X)$;

c. $\mathbf{Var}(X)$.

10. For the random variable of Exercise 6, find:

a. the mean;

b. the variance.

11. A random variable X assumes only the values 0, 1, 2, 3, ..., and the probabilities that X assumes these values are

$$\frac{2}{3}, \frac{2}{3}\frac{1}{3}, \frac{2}{3}\left(\frac{1}{3}\right)^2, \frac{2}{3}\left(\frac{1}{3}\right)^3, \dots,$$

respectively. Find:

a. $\mathbf{E}(X)$;

b. $\mathbf{Var}(X)$.

12. Suppose X is a random variable with generating function

$$f(z) = e^z - e + 2 - z.$$

Find:

a. $P(X = 3)$;

b. $\mathbf{E}(X)$;

c. $\mathbf{Var}(X)$.

13. A random variable takes only the values 3, 4, 5, 6, ..., and the probabilities that it assumes these values are given by

$$P(X = k) = 48/4^k \qquad \text{for } k = 3, 4, 5, \dots.$$

Find the:

a. mean,

b. variance.

14. A random variable takes only the values 2, 4, 6, 8, ..., and the probabilities that it assumes these value are given by

$$P(X = k) = 3/2^k \qquad \text{for } k = 2, 4, 6, \dots.$$

Find the:

a. mean;

b. variance.

15. A random variable X assumes only the values 0, 1, 2, 3, ..., and the probabilities that X assumes these values are

$$1 - \log 1.5, \frac{1}{3}, \frac{1}{3^2}\frac{1}{2}, \frac{1}{3^3}\frac{1}{3}, \frac{1}{3^4}\frac{1}{4}, \dots.$$

a. $\mathbf{E}(X)$;

b. $\mathbf{Var}(X)$.

16. A random variable X has generating function $g(z) = k(e^z + e^{-z})$, where k is a constant. Find:

 a. k;

 b. $E(X)$;

 c. $Var(X)$.

17. Let

$$L = \left(\frac{1}{2} + \frac{1}{2}\frac{1}{4} + \frac{1}{3}\frac{1}{8} + \frac{1}{4}\frac{1}{16} + \frac{1}{5}\frac{1}{32} + \cdots \right)^{-1}.$$

 Let X be a random variable that assumes only the values 1, 2, 3, ... and assumes these values with the probabilities

$$\frac{1}{2}L, \ \frac{1}{2}\frac{1}{4}L, \ \frac{1}{3}\frac{1}{8}L, \ldots,$$

 respectively. Find:

 a. $E(X)$ in terms of L;

 b. $Var(X)$ in terms of L;

 c. L in "simplest form."

18. A coin is tossed repeatedly until it falls the same way twice in a row. Let X be the number of tosses necessary. Find:

 a. $P(X = k)$ for each positive integer k;

 b. the generating function of X;

 c. $E(X)$;

 d. $Var(X)$.

19. A die is thrown repeatedly until it falls "six." If it falls "six" on or before the tenth throw, no money changes hands. If it takes 11 or more tries to get a "six," Alan pays Zeb $1 for each toss beyond the tenth. In this case, Betty pays Zeb $1 for each throw including the first ten. Let X be the amount Alan pays, Y the amount Betty pays, and Z the amount Zeb gets. Find $E(X)$, $E(Y)$, and $E(Z)$.

20. Let the random variable X have generating function f. Show that $[1+f(-1)]/2$ is the probability that X assumes an even value, and $[1 - f(-1)]/2$ is the probability that X assumes an odd value.

21. By applying the result of the last problem, determine under which conditions X is more likely to take an even value than an odd value, when the distribution of X is:

    a. binomial;

    b. Pascal;

    c. geometric;

    d. Poisson.

22. Use generating functions to find the sum of each of the series:

    a. $1 + \dfrac{2}{2} + \dfrac{3}{4} + \dfrac{4}{8} + \dfrac{5}{16} + \cdots$.

    b. $3 + \dfrac{4}{3} + \dfrac{5}{9} + \dfrac{6}{27} + \cdots$.

    c. $\dfrac{1}{4} + \dfrac{2}{8} + \dfrac{3}{16} + \dfrac{4}{32} + \cdots$.

    d. $1 - \dfrac{3}{4} + \dfrac{4}{8} + \dfrac{5}{16} + \cdots$.

    e. $1 + \dfrac{1}{2}\dfrac{1}{2} + \dfrac{1}{3}\dfrac{1}{4} + \dfrac{1}{4}\dfrac{1}{8} + \dfrac{1}{5}\dfrac{1}{16} + \cdots$.

    f. $\dfrac{1}{2} + \dfrac{1}{3}\dfrac{1}{3} + \dfrac{1}{4}\dfrac{1}{9} + \dfrac{1}{5}\dfrac{1}{27} + \cdots$.

    g. $\dfrac{2 \cdot 1}{4} + \dfrac{3 \cdot 2}{8} + \dfrac{4 \cdot 3}{16} + \dfrac{5 \cdot 4}{32} + \cdots$.

8

Random Walks

While this chapter is entitled "Random Walks," random walks will be mentioned by name only in a few paragraphs. Superficially, the chapter seems to be about the story of two characters named Peter and Paul. Actually, what is involved is an abstract idea that may be made concrete in two different ways. The idea is most effectively put into words by discussing Peter and Paul. For this reason, we discuss Peter and Paul. In thinking about the problems we face, a picture in which a "particle" is actually "walking" about is often helpful. We shall explain how to imagine such a picture. A visualization of the moving particle is helpful; a verbal description of it is not. In any case, the Peter-and-Paul approach has historical interest.

8.1   The Probability Peter Wins

The history of Peter and Paul, as opponents in games of chance, starts in France in 1708. (Of course, when two arbitrary names are needed, those two names have been used for a long, long time. For an example involving probability, see Exercise 21 of Chapter Three.) In 1708, Montmort published a book applying probability theory to games of chance. Montmort gave names to his players; he assigned the names by taking them from the list, Pierre, Paul, Jacques, Jean, and Thomas, as he needed them. Thus in the most common situation where the game had two players, the players were called Pierre and Paul. Montmort corresponded with other students of probability theory, and they went along with his choice of names. Almost every probability book written from that time to the present has discussed Pierre and Paul. Those books written in English, have, of course, anglicized the names to Peter and Paul. Occasionally someone has used different names, but we'll go along with the overwhelming precedent.

We digress for a moment to discuss Montmort and his book, *Essay d'analyse sur les jeux de hazard*. This book was the first book on probability theory to be published. In the book, Montmort describes, and computes probabilities for, as many games of chance as he could find. He even discusses a game played by the people he refers to as the "savages of Canada." The second edition of the book, published in 1713, includes an appendix consisting of a number of letters. Among these letters is some of the correspondence between Fermat and Pascal. Montmort wrote for his own amusement and published his book anonymously. In reproducing his correspondence with the Bernoullis, he refers to himself as "M. de M...." We present our brief biography for Montmort here.

Rémond de Montmort

Pierre Rémond de Montmort
(French, 1678–1719)

(That Montmort called himself Rémond de Montmort is inferred from the fact that he reproduces the signature on his letters to the Bernoullis as "R. de M") Montmort was born of a noble family and was destined by his father for the magistrature. "Fatigued" by his studies of law, he travelled to England and Germany. He returned to France in 1699. Shortly thereafter his father died, leaving him a considerable fortune. In 1700 he revisited England and met Newton. In that he year he also obtained an ecclesiastical position at Notre

Dame in Paris; specifically, he obtained a *canonicat*. The nature of that office may be judged by noting that the French word, *canonicat*, is used figuratively to refer to a sinecure. With plenty of money and no duties, Montmort studied mathematics and philosophy. He used his wealth to publish books, some of them about mathematics, that the publishing houses regarded as too risky. He also undertook works of charity, about which he demanded absolute silence. In 1704, he purchased an estate near the village of Montmort. Paying his respects to the most distinguished of his new neighbors, the Duchess of Angoulême, he met her grand-niece, whom he married. In 1715, he made a third visit to England, this time to witness a solar eclipse. While in England, he was elected to the Royal Society. In 1716, he became an associate member of the French academy of sciences; only an associate member, and only that late, because full membership required residence in Paris. In 1719, he died of smallpox, at the age of 41.

Note that *Pierre* Montmort was hardly neutral between *Pierre* and Paul. We still root for Peter and tend to look at things from his point of view. For example, we shall use the letter p, which we have used for the probability of success, for the probability that Peter wins a game; q, corresponding to failure, is the probability Paul wins. Sometimes it is important to be aware that, notation apart, the roles of the players are completely interchangeable. There is a basic symmetry in all of our ideas, if not in our terminology.

Let us explain the words, "random walk." We shall be concerned here only with random walks on the line, that is, one-dimensional random walks. Random walks in higher dimensions are also important. We shall be assuming that Peter and Paul play a series of games. Suppose they bet a fixed amount on each game. Imagine the amount of money that Peter possesses to be continuously displayed graphically on an electric sign. A spot of light moves back and forth along a line to illustrate how Peter is doing. Each time Peter wins, the spot moves one unit to the right; when he loses, it moves one unit to the left. The reader should visualize the sign from time to time, even though we don't explicitly refer to it again. In fact drawing a diagram showing the line and key points, such as the starting point, on it can be very helpful. We should note that both physical and metaphorical random walks occur often in the real world; their mathematical theory has many applications. The reasons we study the present topic here have nothing to do with gambling.

In this chapter we study a certain situation. Throughout the chapter we shall always make the following assumptions. Peter and Paul are going to play a series of games. The probability of Peter's winning any one game is the same as the probability of his winning any other game. Call this probability p, and let q be the probability of Paul's winning any one game. We suppose $p = 1 - q$; thus one player or the other must win in each game; there are no ties. The event of Peter's winning any one game is independent of the event of his winning any other games. Peter and Paul keep score by recording the number of games each has won. To make the situation more vivid, we may suppose a bet is made on each game. In this case, each player bets a certain amount—$1 is simplest, but $1000 is more exciting—on each game. The loser of each game pays the winner the stated amount. We make no overall assumption about how many games are played.

Let us begin by considering a specific example. Suppose Paul is twice as likely to win in any game as Peter is. Thus $p = 1/3$ and $q = 2/3$. To encourage Peter to play, Paul agrees that Peter may decide after each game whether to play another game. Suppose Peter adopts the strategy of waiting until he is one game ahead and then ending the series. Does this ensure that Peter and Paul stop playing with Peter one game ahead? The alternative, in theory at least, is that they continue playing forever without Peter ever being ahead at all. Let us try to compute the probability of each of those two possibilities.

If Peter is to be ahead sooner or later, there must be a first time he is ahead. And there can be only one *first* time he is ahead. Let f_n be the probability that Peter is ahead for the first time immediately after the nth game. The probability that Peter is ever ahead is

$$h = f_1 + f_2 + f_3 + \cdots.$$

We next try to find $f_1, f_2, \ldots$, one at a time. We begin by noting the following: The first time Peter is ahead, he is necessarily ahead by just one game. f_1 is easy. If Peter wins the first game, he is, at that point, ahead by just one game, obviously for the first time. Thus $f_1 = 1/3$.

f_2 is only a little harder. No matter how the first two games turn out, there is no way for Peter to be ahead by just one game immediately after the second game. In short, $f_2 = 0$.

Now we come to f_3. If Peter is to be ahead *for the first time* after the third game, he can not be ahead at the end of the first game. Thus he must lose the first game, and obviously then he must win the next two games. Thus $f_3 = (2/3)(1/3)(1/3) = 2/27$.

f_4 must be zero, like f_2. We may as well generalize. If Peter is ahead by one game, he has won one game more than he has lost. Thus the total number of games must be odd. In short, $f_n = 0$ for all even n.

Next we find f_5. To be ahead for the first time immediately after the fifth game, Peter must win the fourth and fifth games. As before, Peter must lose the first game. Peter must also win just one of the second and third games. His record, with obvious notation, must be LWLWW or LLWWW. The probability of each of these cases is $(1/3)^3(2/3)^2$. Thus $f_5 = 2(1/3)^3(2/3)^2 = 8/243$.

We shall not give the details for f_7 here. There are five possible patterns of wins for Peter: LLLWWWW, LLWLWWW, LLWWLWW, LWLLWWW, and LWLWLWW. Taking that as known, we have $f_7 = 5(1/3)^4(2/3)^3 = 40/2187$.

Let us summarize where we stand. $f_n = 0$ for all even n. For odd n, we have shown:

$$f_1 = 1/3,$$
$$f_3 = (1/3)^2(2/3),$$
$$f_5 = 2(1/3)^3(2/3)^2,$$
$$f_7 = 5(1/3)^4(2/3)^3.$$

Can we predict what would happen if we continued? f_{2k+1} is the probability of Peter's being ahead for the first time immediately after the $(2k + 1)$st game. For this event to happen, Peter must win $k + 1$ of the first $2k + 1$ games and lose k of them. But not any $k + 1$ games will do. f_{2k+1} is $(1/3)^{k+1}(2/3)^k$ multiplied by the number of ways of selecting the $k + 1$ games from the $2k + 1$ so that Peter is never ahead before the $(2k + 1)$st game. The hard part is to find this number of ways, which we shall denote by w_k. We have found $w_0 = 1$, $w_1 = 1$, $w_2 = 2$, and $w_3 = 5$; and our work is getting rapidly harder. Adding a few more terms to the sequence—we're not saying how we got them—we have 1, 1, 2, 5, 14, 42, 132, 429. No general pattern is apparent, and we may as well abandon this approach towards finding h, the probability that Peter is ever ahead.

However, we have accomplished something, besides just getting practice in working with probabilities. The numbers w_k introduced in the last paragraph obviously do not depend on our choice of p, the probability that Peter wins for each game. Thus we have seen that there is a single sequence $w_0, w_1, w_2, \ldots$ of positive integers such that for all p we have...

$$h = w_0 p + w_1 p^2 q + w_2 p^3 q^2 + w_3 p^4 q^3 + \cdots,$$

where h is the probability that Peter is ever ahead.

Now we try a different approach towards finding h. Recall that h may also be described as the probability that Peter is ever ahead by one game. Let t be the probability that Peter is ever ahead by two games. To be ahead by two, Peter must first be ahead by one. At this point, assuming it occurs, we may start keeping score all over again, and again Peter must get to be ahead by one. Thus to be ahead by two, Peter must get to be ahead by one twice in a row. Hence, informally at least, we have $t = h^2$. (The formal details can be supplied; the difficulty is that, in the case where Peter will never be ahead by one, it takes "forever" to find out that that is the case.) We use $t = h^2$ to determine h.

The method used now, and many time hereafter, is to consider the outcome of the first game. If Peter wins the first game, he is then ahead; the probability this happens is p. If Peter loses the first game, he is then one game behind and must, counting from that point, make a net gain of two games; the probability that this happens is qt. Thus we have $h = p + qt = p + qh^2$. The roots of the quadratic equation $h = qh^2 + p$ are

$$h = \frac{1 \pm \sqrt{1 - 4pq}}{2q}.$$

To simplify, we note $1 - 4pq = 1 - 4q(1-q) = 1 - 4q + 4q^2 = (1 - 2q)^2$. Thus the roots of the quadratic equation are

$$\frac{1 + (1 - 2q)}{2q} \quad \text{and} \quad \frac{1 - (1 - 2q)}{2q}$$

For the first of these, we have

$$\frac{1 + (1 - 2q)}{2q} = \frac{2 - 2q}{2q} = \frac{1 - q}{q} = \frac{p}{q}.$$

The second yields

$$\frac{1 - (1 - 2q)}{2q} = \frac{2q}{2q} = 1.$$

Thus, either $h = p/q$ or $h = 1$. While this does not determine h in general, it certainly narrows things down.

Suppose Peter has a greater chance of winning in each game than Paul does; then $p > q$, and hence $p/q > 1$. In this case, p/q is not the probability of anything, and hence $h = 1$. If Peter and Paul have the same chance of winning, then $p = q$, and hence $p/q = 1$. Thus 1 is the only possible value for h in this case. To summarize, if Peter has at least as good a chance in each game as Paul, it is certain that sooner or later Peter will be ahead.

Before determining the value of h for $p < q$, we recall a fact that was established earlier. There is a fixed sequence $w_0, w_1, \ldots$ of positive integers such that

$$h = w_0 p + w_1 p^2 q + w_2 p^3 q^2 + \cdots$$

for every value of p and the corresponding values of q and h. In particular, from the last paragraph with $p = q = 1/2$, we see

$$1 = w_0(1/2) + w_1(1/2)^3 + w_2(1/2)^5 + \cdots.$$

Now consider some p with $p < 1/2$, in other words, with $p < q$. We compare the two infinite series in the last paragraph. Let $a = 1/2 - p$; then $pq = p(1 - p) = (1/2 - a)(1 - 1/2 + a) = (1/2 - a)(1/2 + a) = 1/4 - a^2 < 1/4$. Thus we have

$$
\begin{aligned}
p \quad &< \quad\quad\quad\quad\quad\quad\quad\quad\quad 1/2 \\
p^2 q = p(pq) \quad &< (1/2)(1/4) \quad = (1/2)^3 \\
p^3 q^2 = p(pq)^2 \quad &< (1/2)(1/4)^2 = (1/2)^5 \\
p^4 q^3 = p(pq)^3 \quad &< (1/2)(1/4)^3 = (1/2)^7,
\end{aligned}
$$
etc.

Looking at the last paragraph, we see that each term of the series for h is less than the corresponding term of the series for 1. Thus $h < 1$;

in particular, $h \neq 1$. But we know that either $h = p/q$ or $h = 1$. Thus $h = p/q$. This completes the determination of h in all cases.

We next find, for each positive integer k, the probability that Peter is ever ahead by k games. For $k = 1$, this is the probability h just found. As noted above, the probability that Peter is ever ahead by two games is h^2. Similarly, since to be ahead by k games means that Peter must succeed in getting one game ahead k times in a row, h^k is the probability of Peter's ever being k games ahead. We summarize our results: The probability that Peter is ever ahead by k games is

$$\begin{cases} 1 & \text{if } p \geq q, \\ (p/q)^k & \text{if } p < q. \end{cases}$$

In the discussion just concluded, play continued no matter how far Peter was behind. Suppose a one-dollar bet is made on each game. Then we were considering a situation where Paul was willing to extend Peter unlimited credit. An alternative situation arises when Peter and Paul each start with only so much money and play stops when one of them goes broke. We now study this new situation.

It is clear that one or the other of the players must go broke. Towards seeing why, we first consider the case where $p \geq q$. We saw above that, if play continues long enough, Peter will be one dollar ahead. If play continues from that point, he will get to be a second dollar ahead. Thus eventually, Peter will be ahead by any amount named in advance. Thus Peter will win all Paul's money unless play is stopped by Peter's going broke. If $p < q$, we may interchange the roles of the players and still conclude that one of them must go broke. We may be satisfied to know just that much, but Peter and Paul will want to discuss their individual chances.

Before deriving the general formulas, it is instructive to study some special cases. The method used here is fundamental in this chapter and the next one. The method may also be employed in many cases where the precise circumstances necessary for the formulas to hold do not apply.

In the first problem we work, besides the assumptions already announced, we suppose that Peter starts with $1 and Paul starts with $2. We also suppose $p = q = 1/2$. We are discussing the situation where each player bets $1 on each game and play continues until one of the players goes broke. What is the probability Paul is the one to go broke? We present two ways to work this problem.

In both solutions, we note that Peter must win the first game in order to have a chance of bankrupting Paul. One way to proceed is now to consider the cases, "Peter wins the first two games" and "Peter wins the first game, but loses the second." For brevity, we denote these possibilities by WW and WL. If WW, Paul goes broke. The probability of Paul's going broke because WW occurs is 1/4. After WL, the players each have the same amount of money as they started with; thus they each have the same chance of winning as they did at the outset. Denote the chance that, starting from scratch, Paul goes broke by x. The probability of WL is 1/4. After WL, x is the probability that Paul goes broke. Thus the probability that Paul goes broke with play starting with WL is $(1/4)x$. Combining all this,

$$x = \frac{1}{4} + \frac{1}{4}x.$$

It follows $x = 1/3$; that is, the probability that Paul goes broke is $1/3$.

Now consider the second method of solving our problem. Unless Peter wins the first game, Paul cannot go broke. If Peter does win the first game, after that game, Peter has $2 and Paul $1—just the reverse of the amounts they started with. Thus, at this point, after Peter has won the first game, the probability of Paul's going broke is equal to what was the probability of Peter's going broke when they started. Let x be the probability Paul goes broke, starting from scratch. Then $1 - x$ is the probability of Peter's going broke, also starting from scratch. Therefore $1 - x$ is the probability of Paul's going broke after the first game has been won by Peter. For Paul to go broke, we must first have Peter winning the first game, which has probability 1/2, and then have Paul going broke after that, which has probability $1 - x$. Thus

$$x = \frac{1}{2}(1 - x).$$

Again $x = 1/3$, as we found by the other method.

Now we change the conditions a little. We still suppose Peter begins with $1 and Paul with $2. We now assume $p = 2/3$ and $q = 1/3$. Now what is the probability that Paul goes broke? With symmetry destroyed, the second of the two methods we just used no longer works. The first method gives, by exactly the same reasoning as before,

$$x = \frac{4}{9} + \frac{2}{9}x.$$

Thus $x = 4/7$ is the probability Paul goes broke.

Now we are ready to work the general problem. We still suppose each player bets \$1 on each game and play continues until one of the players goes broke. Since the total amount of money Peter and Paul have between them is fixed throughout our discussion, it is convenient to denote this total by t. Of course, we assume t is a positive integer. If, at any time, Peter has x dollars, then Paul will necessarily have $t - x$ dollars. Peter's chance of being the overall winner clearly depends on how much money he starts with. Let $p_1, p_2, \ldots, p_{t-1}$ be the probabilities that Peter winds up with all the money assuming he starts with $1, 2, \ldots, t - 1$ dollars respectively. Given that there is a time when Peter has i dollars, p_i is the probability that Peter is the overall winner; what happened, if anything, previous to that time is obviously irrelevant to how the game will continue. It is convenient to complete the picture by assigning the values of $p_0 = 0$ and $p_t = 1$.

The first step towards finding the values of the p_i is to see how they are interrelated. Suppose Peter has i dollars. He either wins the next game or loses it. If he wins, the probability is then p_{i+1} that he will go on to be the overall winner. On the other hand, if Peter loses the next game, the probability is then p_{i-1} that he will be the overall winner. Thus we have

$$p_i = pp_{i+1} + qp_{i-1} \qquad \text{for } i = 1, \ldots, t - 1.$$

We also know $p_0 = 0$ and $p_t = 1$. (In formal terms, we have a difference equation with boundary conditions. We shall proceed, however, on an *ad hoc* basis without using the theory of difference equations.) Since $p + q = 1$, we have

$$(p + q)p_i = pp_{i+1} + qp_{i-1}.$$

Thus

$$q(p_i - p_{i-1}) = p(p_{i+1} - p_i).$$

If we set $r = q/p$, the equation

$$p_{i+1} - p_i = r(p_i - p_{i-1}) \qquad (*)$$

may be used to express the other p_i in terms of p_1. Setting $i = 1$ in $(*)$ we have

$$p_2 - p_1 = r(p_1 - 0) = rp_1.$$

Now using $(*)$ with $i = 2$ we have

$$p_3 - p_2 = r(p_2 - p_1) = r^2 p_1.$$

Continuing in this way we see

$$p_4 - p_3 = r^3 p_1,$$

$$\vdots$$

$$p_t - p_{t-1} - r^{t-1} p_1.$$

Adding the sum of the first $j - 1$ of these equations to $p_1 - p_0 = p_1$, we have

$$p_j - p_0 = (p_1 - p_0) + (p_2 - p_1) + \cdots + (p_j - p_{j-1}) = p_1(1 + r + r^2 + \cdots + r^{j-1}).$$

Since $(1 - r)(1 + r + r^2 + \cdots r^{j-1}) = 1 - r^j$ and $p_0 = 0$, we have, unless $r = 1$,

$$p_j = \frac{1 - r^j}{1 - r} p_1. \qquad (**)$$

Now for $j = t$ this gives us

$$p_t = \frac{1 - r^t}{1 - r} p_1.$$

But we know $p_t = 1$; thus

$$p_1 = \frac{1 - r}{1 - r^t}.$$

Returning to $(**)$, we have for all j

$$p_j = \frac{1 - r^j}{1 - r} \frac{1 - r}{1 - r^t} = \frac{1 - r^j}{1 - r^t}.$$

(As noted above, we need $r \neq 1$, that is, $p \neq q$; if $r = 1$, the last equation is meaningless.)

To summarize: Suppose Peter starts with s dollars and Paul starts with $t - s$ dollars. Let p^* be the probability that Peter wins all the money. Then

$$p^* = \frac{1 - r^s}{1 - r^t},$$

where $r = q/p$, unless $p = q$. We state the result for $p = q = 1/2$ here, even though we shall defer the proof. (It is easy to modify the reasoning above to cover this case, but an even easier way to derive the formula will be apparent later.) If $p = q = 1/2$, then

$$p^* = \frac{s}{t}.$$

The situation we have just discussed is often referred to under the name of Gamblers' Ruin. The first printed reference to it occurs in the short treatise of Huygens; we mentioned this treatise in Chapter One. Huygens ends his work with five problems, the last of which is, except for some unimportant extra complications, the special case of what we just did with $p = 5/14$, $t = 24$, and $s = 12$. Huygens does not give any hint as to the method to be used, but he does give an answer. He states that the ratio of the probability that Peter is the overall winner to the probability that Paul is the overall winner is 244140625 to 282429536481. Montmort, in the first edition of his book, presents a way of working the problem. Unfortunately, Montmort, because he was using a French translation of Huygens's treatise instead of the original Latin, worked a different problem from the one Huygens proposed. While the answer Huygens gives is not the correct answer to this different problem, Montmort, apparently using methods well-known to many students, gets the answer "shown in the book." Johann Bernoulli, the brother of Jakob Bernoulli, for whom Bernoulli trials are named, wrote a letter to Montmort pointing out what we just said. (See Chapter Five for biographical data and a table of Bernoullis.) Johann Bernoulli also noted that Montmort had overlooked the fact that the two big numbers in Huygens's answer are 5^{12} and 9^{12}. Montmort agreed that Bernoulli was right on both points. In the second edition of his book, Montmort works the correct problem. He obtains the same system of linear equations that we do. (Since there are 23 unknowns and no subscripts are used, Montmort uses all the letters of the alphabet except a, j, and v as unknowns.) Johann Bernoulli concluded his letter with a postscript indicating he was enclosing some comments from his nephew.

Nikolaus Bernoulli was the nephew of Jakob and Johann. In a letter to Montmort, he gave the general formula we give above. This letter was written almost a year after the letter of Johann Bernoulli referred to in the last paragraph. The general formula also appears, with a proof by mathematical induction, in the book of Jakob Bernoulli. Since this book was published posthumously after editing by Nikolaus Bernoulli, it is far from clear which Bernoulli did what.

Exercises

In those exercises below that mention Peter and Paul, we continue to make our basic assumptions about how those persons gamble. These assumptions are described in the seventh paragraph of this chapter.

1. Peter and Paul bet one dollar each on each game. Peter starts with s dollars and Paul with $t - s$ dollars. They play until one of them is broke. For practice in a certain kind of reasoning, answer the following questions without using the general formulas. Some parts of the exercise require the use of the result of a previous part. What is the probability that Peter wins all the money if:

	$p =$	$s =$	$t =$
a.	1/2	2	4
b.	1/2	1	4
c.	1/2	2	5
d.	1/2	1	5
e.	1/2	2	6
f.	1/2	1	6
g.	1/2	3	6
h.	1/3	1	3

i. 1/3　2　3

j. 1/3　2　4

k. 1/3　1　4

l. 1/3　3　4

2. Peter and Paul bet one dollar each on each game. For this exercise only, we modify our basic assumptions as follows: Peter is nervous on the first game, and the probability of his winning that game is 1/3. Thereafter, Peter and Paul each have probability 1/2 of winning each game. They play until one of them has a net loss of $2. What is the probability Paul is the one with that net loss?

3. To obtain a certain job, a student must pass a certain exam. The exam may be taken many times. If the student passes on the first try, she gets the job. If not, she still gets the job if at some time the number of passing attempts exceeds the number of failures by two. If at any time the number of failures exceeds the number of passes by two, she is not allowed ever to take the exam again. The probability of the student passing on each particular try is 1/2. What is the probability she gets the job?

4. David and Carol play a game as follows: David throws a die, and Carol tosses a coin. If die falls "six," David wins. If the die does not fall "six" and the coin does fall heads, Carol wins. If neither the die falls "six" nor the coin falls heads, the foregoing is to be repeated as many times as necessary to determine a winner. What is the probability that David wins?

5. Three persons, A, B, and C, take turns in throwing a die. They throw in the order A, B, C, A, B, C, A, B, etc., until someone wins. A wins by throwing a "one." B wins by throwing a "one" or a "two." C wins by throwing a "one," a "two," or a "three." Find the probability that each of the players is the winner.

6. Four persons, A, B, C, and D, take turns in tossing a coin. They throw in the order A, B, C, D, A, B, C, D, A, B, etc., until someone gets heads. The one who throws heads wins. Find the probability that each of the players is the winner.

7. Adam and Eve alternately toss a coin.

a. The first one to throw heads wins. If Adam has the first toss, what is the probability that he wins?

**b. Eve must throw heads twice to win, but Adam need throw heads only once to win. If Adam goes first, find the probability he wins. If Eve goes first, find the probability that Adam wins.

**c. Each player needs two heads to win, and Adam goes first. Find the probability that Adam wins.

8. Peter and Paul bet one dollar each on each game. Each is willing to allow the other unlimited credit. Use a calculator to make a table showing, to four decimal places, for each of $p = 1/10, 1/3$, .49, .499, .501, .51, 2/3, 9/10 the probabilities that Peter is ever ahead by $10, by $100, and by $1000.

9. Suppose Peter and Paul bet $1 on each game and Paul starts with $5. For each game, the probability that Peter wins is 1/10. If Paul extends Peter unlimited credit, what is the probability that Peter will eventually have all of Paul's $5?

10. Repeat the last exercise assuming, for each game, the probability Peter wins is .499.

11. Peter needs $100 for a special purpose; he will stop gambling when he gets it. Peter and Paul bet $10 on each game. If $p = .48$ and Paul extends Peter unlimited credit, what is the probability that Peter gets the $100?

12. Peter has probability 2/3 of winning each game. Peter and Paul bet $1 on each game. If Peter starts with $3 and Paul with $5, what is the probability Paul goes broke before Peter is broke?

13. Peter has probability 1/4 of winning each game. Peter and Paul each bet $100 on each game. They each start with $400 and play until one of them goes broke. What is the probability that Paul goes broke?

14. Peter and Paul each bet $1 on each game, and each player starts with $10. Peter has probability 1/3 of winning in each game. What is the probability that Peter is $3 ahead at some time before he is $7 behind?

15. Peter and Paul each have probability 1/2 of winning in each game. They bet $10 each on each game. What is the probability that Peter is $100 ahead at some time before he is $50 behind?

16. Peter has probability 2/3 of winning in each game. Peter and Paul each bet $100 on each game. Peter starts with $200 and Paul with $600. They play until one of them goes broke. What is the probability that Peter goes broke?

17. Peter starts with $10,000 and Paul with $1,000. They bet $100 each on each game. In each game, each player has the same chance of winning. What is the probability that Peter goes broke?

18. Peter and Paul each start with $32 and $p = .6$. They bet $1 each on each game. Use a calculator to find the approximate probability that Peter bankrupts Paul. [Note: If your calculator does not have a button for raising numbers to arbitrary powers, you can find the 32nd power of a number by squaring five times in a row; since we have $(x^2)^2 = x^4$, $(x^4)^2 = x^8$, $(x^8)^2 = x^{16}$, $(x^{16})^2 = x^{32}$. More generally, the (2^n)th power may be found by squaring n times.]

19. Do the last exercise modified so that Peter starts with $8 and Paul with $56.

20. Do Exercise 18 modified to provide $p = .51$, Peter starts with $8, and Paul with $56.

21. Do Exercise 18 modified to provide $p = .501$, Peter starts with $256, and Paul with $768.

22. Peter starts with $2048 ($2048 = 2^{11}$) and $p = .501$. Paul starts with billions. Use a calculator to find the approximate probability that Peter bankrupts Paul, assuming each player bets $1 on each game. (See the note to Exercise 18.)

23. Show that, for each positive integer k, there is a sequence $a_0, a_1, a_2, \ldots$ of nonnegative integers such that the probability Peter is ever k games ahead is

$$a_0 p^k + a_1 p^{k+1} q + a_2 p^{k+2} q^2 + a_3 p^{k+3} q^3 + \cdots,$$

for all values of p.

**24. If $w_0, w_1, \ldots$ are the numbers defined in the beginning of this chapter, show

$$w_n = \frac{1}{n+1}\binom{2n}{n}.$$

8.2   The Duration of Play

So far we have considered only whether certain things occur if Peter, Paul, and we all wait long enough. It is often important to have some idea of how long it is necessary to wait. For example, if play continues until someone goes broke, we should study the random variable Y, the number of games until that happens. Alternatively, given k and $p \geq q$, Y can be the number of games before Peter is k games ahead; we know that will happen sooner or later. On the other hand, if $p < q$, it makes no sense to consider how many games it takes for Peter to be k games ahead, since that may never happen. When play is sure to end sometime, it is interesting to consider $\mathbf{E}(Y)$, the average number of games to be played. A preliminary question is whether $\mathbf{E}(Y)$ is defined. In other words, does the infinite series that defines $\mathbf{E}(Y)$ converge? We leave that question for the exercises (see Exercises 51 and 52). We begin our study of how long play continues by returning to some examples considered earlier.

As before, besides our general assumptions, we suppose Peter starts with \$1 and Paul with \$2. We also suppose each player bets \$1 on each game and play continues until someone is broke. How long does that take, on the average? (As we just said, here we simply assume that there is a finite average, leaving the proof for exercises.) First we do the problem with $p = q = 1/2$. The same two methods already used to find the probability of Paul's going broke apply again. We use the same notation, with W and L, as before. For example, WL means Peter wins the first game and loses the second; the results of the remaining games are not specified. We describe the two methods in turn.

In the first method, we distinguish the cases L, WL, WW. Note that just one of these cases must in fact arise. The probability of L, Peter losing the first game, is 1/2. In this case, play stops after a single game. The probability of WW is 1/4; in this case, two games settle the matter. The probability of WL is 1/4. However, after WL, more games must be played. We don't know how many additional games are necessary before someone goes broke. However, note that after WL each player has the same number of dollars he started with. Thus, on the average, after WL as many games remain to be played as, on the average, were necessary starting from scratch; the two games completed have accomplished nothing. Denote the expected number of games to be played, starting from the beginning, by y; we are trying to find y. After WL, the expected number of games still to be played is y, as we just said. Thus, assuming WL, the total number of games to be played is, on the average, $y + 2$. The 2 represents the number of games that were "wasted" by WL; WL took two games to get back where we started. Formally the following reasoning uses conditional expectation, but informally it is intuitively obvious. We combine the conclusions above into an equation, marking the terms with symbols indicating the situation covered in each of them:

$$\overset{\text{L}}{y = \frac{1}{2} \cdot 1} + \overset{\text{WW}}{\frac{1}{4} \cdot 2} + \overset{\text{WL}}{\frac{1}{4}(y + 2)}.$$

Solving for y we have $y = 2$, our answer.

Now we try the other way of getting this answer. We distinguish only the cases W and L, according to how Peter fares on the first game. If L, it's all over in one game. If W, after one game the players have just exchanged the amounts of money they have. Thus, in this case, after the first game, as many games will still be needed, on the average, to bankrupt one of the players as were needed, on the average, at the beginning. Adding the one "wasted" game, and using y as before, $y + 1$ is the expected number of games to be played if Peter wins the first game. We have the equation

$$y = \frac{1}{2} \cdot 1 + \frac{1}{2}(y + 1).$$

Again we find the answer $y = 2$.

Next we modify the conditions and set $p = 2/3$, $q = 1/3$. The second method no longer works since we have lost the symmetry it needed. The first method, by the same reasoning we just used, gives us

$$\overset{\text{L}}{y = \frac{1}{3} \cdot 1} + \overset{\text{WW}}{\frac{4}{9} \cdot 2} + \overset{\text{WL}}{\frac{2}{9}(y + 2)}.$$

Thus $y = 15/7$, and, on the average, it takes $15/7$ games for one of the players to go bankrupt.

As a first step towards deriving a general formula that covers the example just completed, we consider a much more general situation.

We suppose that Peter and Paul agree to bet $1 each on each of certain games. Whether or not a bet is made on a particular game may be a matter of chance, perhaps depending on the outcome of previous games. However, whether a bet is made on a game and the outcome of *that* game are to be independent events. In other words, given that a bet is made on a certain game, the conditional probability that Peter wins that game is p. Let X be the net amount that Peter wins overall and Y be the number of games on which bets are made. Whether $E(Y)$ is defined depends on the agreement between Peter and Paul; the average number of games on which bets will be made need not be finite. We shall determine the relationship between $E(X)$ and $E(Y)$ on the assumption that $E(Y)$ is defined.

We begin by defining, for each of $n = 1, 2, \ldots$, random variables X_n and Y_n as follows:

$$X_n = 1, \quad Y_n = 1 \qquad \text{if Peter bets on and wins the } n\text{th game.}$$
$$X_n = -1, \quad Y_n = 1 \qquad \text{if Peter bets on and loses the } n\text{th game.}$$
$$X_n = 0, \quad Y_n = 0 \qquad \text{if no bet is made on the } n\text{th game.}$$

Thus $X = X_1 + X_2 + \cdots$ and $Y = Y_1 + Y_2 + \cdots$. Let c_n be the probability that a bet is made on the nth game. Then we have

$$E(X_n) = 1 \cdot P(X_n = 1) + (-1)P(X_n = -1) = c_n p - c_n q = c_n(p - q).$$
$$E(Y_n) = P(Y_n = 1) = c_n.$$

Thus, for all n, we have $E(X_n) = E(Y_n)(p - q)$. Under the assumption that $E(Y)$ is defined, it can be shown that $E(X) = E(X_1) + E(X_2) + \cdots$ and $E(Y) = E(Y_1) + E(Y_2) + \cdots$, even though there are infinitely many terms in each sum. Hence we can conclude that

$$E(X) = E(Y)(p - q).$$

We make several applications of the formula just derived. Return to the situation where Peter starts with s dollars, Paul starts with $t - s$ dollars, and they make one-dollar bets on every game until someone goes broke. Y is then the number of games it takes before someone goes broke. We can easily find $E(X)$. Let p^* be the probability that Peter winds up with all of the money; we have a formula for p^* already. Then $1 - p^*$ is the probability that Paul is the overall winner. If Peter gets all Paul's money, Peter finishes $t - s$ dollars ahead.

Otherwise, Peter loses all his s dollars. Thus

$$\mathbf{E}(X) = p^*(t - s) + (1 - p^*)(-s) = p^*t - s.$$

It can be shown, using Exercise 49, that $\mathbf{E}(Y)$ is defined in this case. We have then, assuming $p \neq q$,

$$\mathbf{E}(Y) = \frac{p^*t - s}{p - q}.$$

This formula may be used to find $\mathbf{E}(Y)$ when $p \neq q$.

We next consider the case $p = q$. In this case, $\mathbf{E}(X) = \mathbf{E}(Y)(p - q)$ becomes simply $\mathbf{E}(X) = 0$. That won't help us find $\mathbf{E}(Y)$, but we can now derive the formula for p^*, stated above without proof. [We may as well at least announce $\mathbf{E}(Y) = s(t - s)$, even though we leave the proof for the Appendix to this chapter.] We have, as in the last paragraph, $0 = \mathbf{E}(X) = p^*t - s$. Thus, $p^* = s/t$, as claimed.

We pause briefly for an historical sidelight. DeMoivre, whose biography appears in Chapter Six, found an ingenious way to derive the formula for p^* in the case where $p \neq q$. He found a way to use the same simple method we just used for the case $p = q$. See Exercise 51 for the details.

Now we return to the situation considered first. We suppose Paul extends Peter unlimited credit and they continue betting until Peter is k dollars ahead. To be sure that Peter will eventually be k dollars ahead, we need $p \geq q$. Since play stops when Peter is k dollars ahead, $X = k$, no matter how the individual games turn out. For $p > q$, it can be shown that $\mathbf{E}(Y)$ is defined (see Exercise 52). Thus, for $p > q$, we have $k = \mathbf{E}(X) = \mathbf{E}(Y)(p - q)$ and can conclude

$$\mathbf{E}(Y) = \frac{k}{p - q};$$

in other words, it takes $k/(p - q)$ games on the average before Peter is k games ahead. If $p = q$, the formula $\mathbf{E}(X) = \mathbf{E}(Y)(p - q)$ yields $k = 0 \cdot \mathbf{E}(Y) = 0$, which is impossible. The formula fails because $\mathbf{E}(Y)$ is not defined. The original definition of expected value, we recall, sometimes involves an infinite series. When the series diverges, the expected value is undefined. We have just shown that Y has no expected value; in other words, we have shown that a certain series diverges. Informally, if $p = q$, it is certain that Peter will sooner or later be ahead by k games, but on the average it takes "forever" before he does this. It almost makes sense to claim that the formula

$$\mathbf{E}(Y) = \frac{1}{p - q}$$

works whenever $p \geq q$, on the grounds that it gives "infinity" when $p = q$.

The following tables summarize the formulas of this chapter:

Unlimited Credit

	$p < q$	$p = q$	$p > q$
Probability Peter is ever ahead by k	$\left(\dfrac{p}{q}\right)^k$	1	1
Average number of games before Peter is ahead by k	—	—	$\dfrac{k}{p - q}$

No Credit

Peter starts with s. Paul starts with $t - s$.
Play continues until someone is broke.
$r = q/p$.

	$p \neq q$	$p = q$
Probability Paul goes broke	$p^* = \dfrac{1 - r^s}{1 - r^t}$	s/t
Number of games until someone is broke	$\dfrac{p^* t - s}{p - q}$	$s(t - s)$

Exercises

25. Reconsider each of the situations described in Exercise 1. Again for practice in a certain kind of reasoning, answer the following question without using the general formulas. What is the

expected number of games before one of the gamblers loses all his money?

26. Cain and Abel are gambling on a series of games. Each starts with \$2. They bet one dollar each on each game. The one who is ahead becomes overconfident and is more likely to lose. More precisely:

If Cain is ahead by	Probability Cain wins	Probability Abel wins
−1	2/3	1/3
0	1/2	1/2
1	1/3	2/3

They play until one of them goes broke. What is the expected value of the number of games they play?

27. In the situation described in Exercise 2, find the expected number of games Peter and Paul play.

28. In the situation of Exercise 3, how many times does the student take the exam, on the average.

29. In the situation of Exercise 5, how many times is the die thrown, on the average.

**30. a. How many tosses are necessary, on the average, to determine the winner in each of the situations of Exercise 7b?

   b. Repeat part a for Exercise 7c.

31. Peter and Paul bet one dollar each on each game. Paul starts with \$10. Paul extends Peter unlimited credit. If $p = 2/3$, what is the expected number of games before Paul goes broke?

32. Under each of the sets of circumstances of Exercise 8, find the expected number of games before Peter is ahead by the amount specified; say "not applicable" where appropriate.

In Exercises 33–42, find the expected number of games Peter and Paul play in the situation of the indicated exercise.

33. (Instructions above) Exercise 12.

34. (Instructions above) Exercise 13.

35. (Instructions above) Exercise 14.

36. (Instructions above) Exercise 15.

37. (Instructions above) Exercise 16.

38. (Instructions above) Exercise 17.

39. (Instructions above) Exercise 18.

40. (Instructions above) Exercise 19.

41. (Instructions above) Exercise 20.

42. (Instructions above) Exercise 21.

43. If a computer simulates the series of games in Exercise 22 at the rate of one million games per second, how long will it take to finish the simulation, assuming "billions" means "two billion"?

44. Mr. Lucky has $1000 with him. He plans to make a series of independent bets at even money at a casino. The probability of Mr. Lucky's winning any one bet is .49. He plans to continue until either he is $100 ahead or he loses the $1000. Mr. Lucky's bets will all be the same size, either $1, $5, $10, $50, or $100. Use a calculator to estimate the following:

   a. For each size of bet, find the probability that Mr. Lucky succeeds in winning the $100.

   b. For each size of bet, how long does it take, on the average, to settle the matter?

45. A coin is tossed repeatedly until it falls heads twice in a row. Find the expected number of tosses necessary.

46. Do the last problem for heads three times in a row.

47. A die is thrown repeatedly until it falls "six" twice in a row. Find the expected number of throws necessary.

**48. Fill in the details of DeMoivre's derivation of the formula for p^* when $p \neq q$: DeMoivre assumed that Peter and Paul played for "counters" of unequal value. In detail, let $M_1, M_2, \ldots, M_t$ be counters, with M_i worth r^i dollars. Suppose Peter starts with $M_1, \ldots, M_s$ and Paul with the other counters. On each game, Peter bets the most valuable one of his counters while Paul bets the least valuable one of his. Peter and Paul play until someone has all the counters. Show that each bet is fair and assume, as DeMoivre did, that therefore the overall situation is fair. Now proceed as we did for $p = q$.

**49. Consider Peter and Paul with $p > 1/2$. Assume play continues until Peter is ahead. Let Y be the number of games until that happens. Show that, with $w_0, w_1, w_2, \ldots$ as early in the chapter,

$$\mathbf{E}(Y) = w_0 p + 3w_1 p^2 q + 5w_2 p^3 q^2 + 7w_3 p^4 q^3 + \cdots.$$

Show that this series converges by comparing it to

$$w_0(1/2) + w_1(1/2)^3 + w_2(1/2)^5 + w_3(1/2)^7 + \cdots;$$

we have already shown that the latter series converges to 1.

**50. Consider Peter and Paul with $p > 1/2$. Assume play continues until Peter is k games ahead. Let Y be the number of games until that happens. Show that $\mathbf{E}(Y)$ is finite by combining the ideas of the last exercise with those of Exercise 23.

Appendix

We now prove the formula $s(t - s)$ announced above. The circumstances, besides the general assumptions of this chapter, are as follows: Peter starts with s dollars and Paul with $t - s$. They each bet a dollar on each game, and play continues until someone goes broke. Also, each player is as likely to win in each game as his opponent; i.e., $p = q = 1/2$. We seek to show that it requires an average of $s(t - s)$ games before one player has all the money.

Barring trivialities, we suppose $s \geq 1$ and $t - s \geq 1$; if $s = t - s = 1$, it obviously takes just one game to bankrupt one of the players. Thus the formula gives the correct value in this case. We go on to check the formula for $t = 3, 4, 5, \ldots$ *in turn.* (This process of basing the proof for each value of t on having established the formula for lower values of t is called mathematical induction.) The proof for $t = 3$ will use the formula for $t = 2$; the proof for $t = 4$ will use the formula for $t \leq 3$, etc. In other words, we show, for each particular value of t, that the formula holds on the assumption that it holds whenever Peter and Paul start with fewer than t dollars between them.

We need to distinguish three cases according to whether $s < t/2$, $s > t/2$, or $s = t/2$. First suppose $s < t/2$. Then, since Peter starts with less than half the money, Paul starts with more than half. Thus, sooner or later, either Peter will go broke or Paul will have lost all but s of his dollars. How long does that take?

s	$t - 2s$	s

We are concerned with either Peter losing s dollars or Paul losing $t - 2s$ dollars, whichever comes first. Since $s + (t - 2s) < t$, by our assumption about what happens when fewer than t dollars are involved, it takes $s(t - 2s)$ games, on the average, for this situation to arise. If at this point Peter is broke, play is over. If Paul has s dollars left, additional games are necessary. The probability of Peter winning $t - 2s$ dollars before Paul wins s dollars is

$$\frac{s}{s + t - 2s} = \frac{s}{t - s}.$$

Thus that is the probability that additional games are necessary. How many additional games will be needed? We are discussing reaching a situation where Paul has s dollars and Peter $t - s$, just the reverse of the way they started. Thus as many additional games are needed, on the average, as were needed, on the average, overall when play started. Denote this number of games by x. We have

$$x = s(t - 2s) + \frac{s}{t - s}x.$$

Solving for x, we have $x = s(t - s)$ in the case $s < t/2$.

Now we cover the other cases. The case $s > t/2$ is the same as the first case with the roles of the players interchanged. Thus the formula holds for $s > t/2$ also. Finally, suppose $s = t/2$. Then each player starts with s dollars. After one game, one player has $s - 1$ dollars and the other $s + 1$. At that point, by the cases already covered, $(s - 1)(s + 1)$ games remain to be played, on the average. The total expected number of games needed is thus

$$1 + (s - 1)(s + 1) = 1 + s^2 - 1 = s^2 = s(t - s),$$

since $t = 2s$. This completes the proof.

9 Markov Chains

9.1 What Is a Markov Chain?

In this chapter we generalize the situation considered in the last chapter. Recall that in the last chapter Peter and Paul were gambling. At any time during their play, their finances were in a certain condition or state. The word state is the one usually used here. Thus we had a "system" that could be in any one of a number of states. In fact, it moved from state to state as time went on. The motion was in discrete steps; each step corresponded to one game. We shall retain those concepts from the last chapter. We assumed that the amount bet was fixed at one dollar. Thus at each step Peter either gained a dollar or lost a dollar. We now want to generalize and allow the possibility of change in a single step from any state to any state.

Before we actually make the generalization, we want to pick out a very important feature of the gambling situation; this feature we shall retain. Suppose that at a certain time we know that Peter and Paul each have a certain amount of money. The course of their fortunes from that time on depends only on how they stand at that time. What happened earlier—in fact, whether anything at all happened earlier—is irrelevant except insofar as it determined how Peter and Paul stand at the time in question. It only matters where they are, not how they got there.

Now we are ready to describe our basic assumptions for this chapter. We have something that can be in any one of a number of "states." The last sentence is intentionally vague. We are trying to describe a concept that covers a wide variety of examples—some such examples may be found in the exercises. It would be easy to be very specific; each exercise does that. But then we would be trading broad applicability for useless concreteness. On the other hand, we could be abstract; we could, for example, require each state to be a positive integer. Instead we simply say that the states are to be designated $E_1, E_2, \ldots, E_s$. Note particularly that, for simplicity in this elementary treatment, we assume that there are only finitely many states.

What is it that moves from state to state? The basic idea is that the whole system we are studying changes from one condition to another. Sometimes a physical object actually moves, but frequently that is not the case. We can say, for example, "The system is in state E_3." Or we can say, "We have state E_3." Often we involve ourselves and say, "We are in state E_3." All these statements have the same meaning.

One feature of random walks that we definitely want to retain is the following: Taking it as known that the conditions are such-and-such at a given time, what happens next is not affected in any way by when that time is. We restate this by using, in a rather informal way, the conditional probability notation $P(A|B)$:

P(system is in E_j at a certain time | system was in E_i one stage earlier)

is independent of what time is the "certain time." Before saying even more, let us simplify the language by regarding the "certain time" as now. In other words, we view the system as having wandered around in a way of which we have records in the past, as being somewhere now, and as having an unknown future ahead of it. A certain key property of the situation we are now studying is called the *Markov property*:

> P(system is now in E_j | system was in E_i one step ago)
> $= P$(system is now in E_j | system was in E_i one step ago
> and any additional information about the past).

The point is that we get the same value for this probability no matter when "now" is or what the "additional information" is. Of course, the past may affect the future, but only by acting through the present. As long as we are where we are, it doesn't matter how we got there.

At this point, we have described the properties of a Markov chain. Let us just agree that when we speak of a Markov chain, we are saying that we are considering a system along the lines described above. It doesn't really help to select some particular object, such as the transition matrix to be described shortly, and say that it *is* the Markov chain.

In a moment, we'll give the short biography of the Markov for whom Markov chains are named. However, we should first explicitly state that Markov did invent Markov chains. Markov chains formed only a small part of Markov's work on probability. And probability formed only a part of Markov's mathematical work.

A.A. Markov

Andrey Andreyevich Markov (Russian, 1856–1922)

Markov was the son of a middle-level civil servant. He used crutches until he reached the age of ten. As a student, his work was rather poor except in the area of mathematics. When he reached an appropriate age, he entered the University of St. Petersburg, where he remained for his entire career. There he was a student of Chebyshev, joining the circle of mathematicians whose work grew out of the work of Chebyshev. As a teacher, Markov presented lectures that were difficult to understand; he did not bother with the order in which he wrote equations on the blackboard. In 1905, after 25 years of teaching, Markov retired in order to make way for younger mathematicians. However he continued to teach the probability course. In political matters, Markov protested against the authorities to an extent that would not have been tolerated except that he was judged a harmless academic. When the Czar intervened to prevent the author Maxim Gorky (1868–1936) from entering the St. Petersburg

Academy, to which Gorky had just been elected, Markov was particularly active in protesting. In 1913, when the authorities organized a celebration of the 300th anniversary of the Romanovs, Markov organized a celebration of the 200th anniversary of Bernoulli's Law of Large Numbers.

Suppose we have a Markov chain. Then obviously we consider the probability of going from a state E_i to a state E_j. But we first must clarify what we are talking about. We mean the conditional probability, given being in E_i, of going from there to E_j. The start in E_i is taken as given. There still are two possible meanings for the probability of going from E_i to E_j. We may mean the conditional probability given being in E_i at one stage of being in E_j at the next stage; this probability is denoted by p_{ij}. On the other hand, we may mean the conditional probability given being in E_i of being in E_j sooner or later thereafter; this probability is denoted by h_{ij}.

The usual way to specify a Markov chain is to give the values of the p_{ij} for all i and j. These numbers are called the *transition probabilities*. One way to describe them is to list them all in a table. We now adopt a certain convention as to how we shall arrange such a table. (Unfortunately, there are two reasonable arrangements possible. Some people use one and some the other. All we can say is what we shall do.) In the ith horizontal row of our table we put the probabilities of going from E_i to each of the other states. In the jth vertical column, we put the probabilities of coming to E_j from each of the states. In other words, we put p_{ij} in the ith row, jth column. Before giving an example, we state that the square array of numbers just described is called the *transition matrix* of the Markov chain.

The substance of the last paragraph is perhaps best presented by considering an example. Suppose a Markov chain has the transition matrix

$$\begin{bmatrix} 1/2 & 1/4 & 1/6 & 1/12 \\ 1/3 & 1/3 & 0 & 1/3 \\ 0 & 1 & 0 & 0 \\ 1/5 & 1/5 & 1/5 & 2/5 \end{bmatrix}$$

The 1/4 in the matrix is in the first row, second column. Thus $p_{12} = 1/4$. If we start in, or at any time are in, state E_1, then 1/4 is the probability of being in E_2 one step thereafter. Also, $2/5 = p_{44}$ is the probability that after being in E_4 at any one stage we shall also be there one step later. Since $p_{23} = 0$, it is impossible for a visit to E_2 to be followed immediately by a visit to E_3. Look at the third row. If we are in E_3 at any stage, we necessarily will be in E_2 at the next stage. Observe that in the matrix the total of the numbers in each row is 1; from a certain state we go in the next step to just one other state. On the other hand, the columns have no particular number as total.

A question of some importance is whether it is possible to get from state E_i to state E_j. We denote by $\mathcal{R}_i$ the set of those states that there is a positive probability of reaching sooner or later after being in E_i. Thus, $E_j \in \mathcal{R}_i$ means that it is possible to go from E_i to E_j, but not necessarily directly. We have $E_j \in \mathcal{R}_i$ if and only if $h_{ij} > 0$. A special case requires a little extra care as to language. To say that $E_i \in \mathcal{R}_i$ is to say that it is possible that we find ourselves in E_i *again* after having been there once.

It is very easy to determine $\mathcal{R}_i$ for each i. We fix our attention on one i. First we list, in any order, the states that can be reached from E_i in one step. (Recall that we do not know in advance whether $E_i \in \mathcal{R}_i$. We treat E_i here like any other state.) Then we consider the first state on our list. We add to the end of the list all states *not already listed* that can be reached from this first state in one step. Then we check off this first state as having been attended to. Now we consider the next state on the list and treat it just like the first state. This process of checking off steps one at a time from the beginning of the list while adding new states at the end is to be repeated until every state on the list is checked. Eventually all states on the list will be checked since there are only finitely many states in all. Now we have a list of states that can be reached from E_i in one or more steps. Every state that can be reached from a listed state is itself on the list. In short, the states on the list are all of those that can be reached from E_i; in other words, they constitute $\mathcal{R}_i$.

Those states E_i for which $h_{ii} = 1$ differ in a very important way from those for which $h_{ii} < 1$. Suppose $h_{ii} = 1$ and we do reach E_i at some stage. To say $h_{ii} = 1$ is to say we must necessarily return to E_i at a later stage. Then, having returned, we are in E_i, and thus we must again return to E_i at a still later stage. In fact, we return to E_i again and again forever. Thus, if we reach E_i at all, we must reach E_i infinitely many times. Now suppose, on the other hand, that $h_{ii} < 1$ and that we reach E_i at some stage. Then there is a positive probability that we shall never be in E_i again. If we do get back, there is again a chance of leaving for good; and so on. It is clear that we shall eventually pay a last visit to E_i and never come back. Thus E_i can be reached only finitely many times. The following true statement is rather surprising: Assuming a particular state E_i will be reached, whether we shall be in E_i finitely many times or infinitely many times is not a matter of chance. Which alternative occurs is predetermined by the structure of the Markov chain. We call the states E_i for which $h_{ii} = 1$ *recurrent*. If $h_{ii} < 1$, we call E_i *transient*.

We next give a method for determining whether a state is transient or recurrent. Suppose E_i is the state we seek to classify. There are two mutually exclusive possibilities:

1. For every $E_j \in \mathcal{R}_i$, we have $E_i \in \mathcal{R}_j$.

2. There is an $E_j \in \mathcal{R}_i$ such that $E_i \notin \mathcal{R}_j$.

Suppose the second possibility occurs. Then from E_i it is possible to get to a state E_j from which return to E_i is impossible. In this case, given sufficiently many visits to E_i, we would eventually go from E_i to E_j and never thereafter return to E_i. Thus there would be only finitely many visits to E_i. It follows that E_i is transient. Suppose, on the other hand, condition 1 above holds. We reason somewhat informally as follows: If we start in E_i, we are always in states from which a return to E_i is possible. For each state we can get to, there is a positive probability of returning to E_i in some particular number of steps. Since there are only finitely many states that we can get to, we may let p be the smallest of these probabilities and N be the largest of these numbers of steps. Then, wherever we get starting from E_i, there is always a probability of at least p of getting back to E_i in N or fewer steps. This insures, since we keep making repeated tries to return, that sooner or later we shall return. Thus, $h_{ii} = 1$ and E_i is recurrent. To summarize: If there is an $E_j \in \mathcal{R}_i$ with $E_i \notin \mathcal{R}_j$, then E_i is transient. If for every $E_j \in \mathcal{R}_i$ we have $E_i \in \mathcal{R}_j$, then E_i is recurrent.

It is impossible to go, in one or more steps, from a recurrent state to a transient state. We begin proving this last statement by supposing the contrary; in other words, we suppose E_i is recurrent,

E_j is transient, and $E_j \in \mathcal{R}_i$. Since E_j is transient, there is an $E_k \in \mathcal{R}_j$ such that E_j cannot be reached from E_k. But we can go from E_i to E_j and then to E_k. Thus, since E_i is recurrent, we can get from E_k to E_i. Then we can go on to E_j, thus reaching E_j from E_k. This contradiction completes the proof.

What happens in the long run? The transient states can occur only finitely many times each. Thus sooner or later we must reach a recurrent state. Thereafter we can only be in recurrent states.

An extreme case of a recurrent state deserves special mention. If $p_{ii} = 1$, we call E_i *absorbing*. If we reach such an E_i, we shall be there at every stage thereafter. Obviously, an absorbing state is recurrent. We determine which states are absorbing simply by looking at all the numbers p_{ii}; these numbers appear on the diagonal of the transition matrix.

Exercises

1. A museum owns a certain three paintings by Renoir, two by Cézanne, and one by Monet. It has room to display only one of these paintings. Therefore the painting on display is changed once a month. At that time, the painting on display is replaced by a randomly chosen one of the other five paintings. Let E_1 be "A Renoir is on display," E_2 be "A Cézanne is on display," and E_3 be "A Monet is on display." Find the transition matrix for the Markov chain just described.

2. Do the last problem modified as follows: The next painting to be displayed is randomly chosen from among those paintings by different artists from the painting being replaced.

3. Brands A, B, C, and D of detergent are placed in that order on the supermarket shelf. A customer either buys again the same brand as last purchased or changes to a brand displayed next to it on the shelf; if two brands are adjacent to the old brand, and a change of brands is made, a random choice is made between those two. The probability of buying Brand A given that Brand A was the one purchased last is .9; the corresponding probabilities for Brands B,

C, and D are .9, .8, and .9, respectively. Find the transition matrix for the Markov chain involved.

4. The floor plan of a portion of a museum is shown in the diagram below.

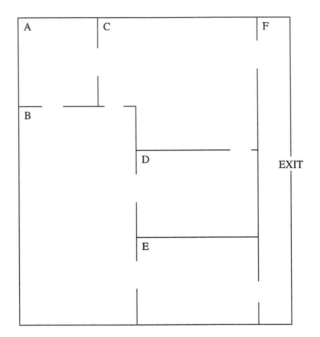

Mr. Hadd E. Nuff wants to leave the museum, but he has become flustered. He is wandering from room to room at random; he is as likely to leave a room by any one door as by any other door. However, when he reaches Room F, he can see the exit, and thus he will definitely go that way and not return. Find the transition matrix for the path taken by Mr. Nuff; use one state for each room and one for "out of the area."

5. A sales representative divides her time among six cities, located as shown in the sketch map below. The map also shows which pairs of these cities have scheduled air service; such pairs are connected by a line segment marked with the number of flights per day in each direction. The representative, for reasons that we do not explain, chooses a flight at random from those going from the city she is in to one of the other five cities; after completing

her business in the latter city, she repeats this process. Find the transition matrix of the Markov chain involved here.

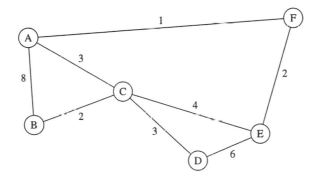

6. There are three slot machines in a row. A man chooses which machine to play according to the following rule: Whenever a machine pays him anything, he plays that machine again. Otherwise he changes to another machine adjacent to the one he was playing; if he was playing the middle machine, he chooses at random from the other two machines. The machines pay off, in order from left to right, 10%, 10%, and 5% of the time. Find the transition matrix.

7. A certain company classifies its credit accounts as "current," "inactive," or "closed." The status of any account for a calendar month is determined on the first day of that month. An account that is current now has probability .8 of still being current a month from now; otherwise it will be inactive. An account that now is inactive, but has been current at some time during the past three months, is as likely to be current as to remain inactive a month from now; it will not be closed. An account that is now inactive and was last current four months ago has probability .2 of being current next month; otherwise it remains inactive. An account that is now inactive for the fifth consecutive month has probability .1 of being current next month; otherwise it will be classified as closed next month. A closed account remains permanently closed. Find the transition matrix of the Markov chain involved here. (Note: The hard part is determining what the states are.)

8. For each of the following transition matrices, determine $\mathcal{R}_i$ for all i. Which states are recurrent, which are transient, and which are absorbing?

a. $\begin{bmatrix} 1/3 & 0 & 0 & 1/3 & 1/3 & 0 & 0 \\ 0 & 1/2 & 0 & 1/2 & 0 & 0 & 0 \\ 0 & 0 & 1 & 0 & 0 & 0 & 0 \\ 1/2 & 0 & 0 & 1/2 & 0 & 0 & 0 \\ 1/2 & 0 & 0 & 1/2 & 0 & 0 & 0 \\ 0 & 0 & 0 & 0 & 0 & 1 & 0 \\ 0 & 1/3 & 1/3 & 0 & 1/3 & 0 & 0 \end{bmatrix}$

b. $\begin{bmatrix} 0 & 0 & 0 & 0 & 1/2 & 0 & 1/2 & 0 \\ 0 & 1/3 & 0 & 0 & 0 & 2/3 & 0 & 0 \\ 0 & 1/3 & 0 & 0 & 1/3 & 0 & 0 & 1/3 \\ 0 & 0 & 0 & 1 & 0 & 0 & 0 & 0 \\ 1/2 & 0 & 0 & 0 & 1/2 & 0 & 0 & 0 \\ 0 & 1/2 & 0 & 0 & 0 & 1/2 & 0 & 0 \\ 1/3 & 0 & 0 & 0 & 1/3 & 0 & 1/3 & 0 \\ 0 & 0 & 1/2 & 1/2 & 0 & 0 & 0 & 0 \end{bmatrix}$

c. $\begin{bmatrix} 0 & 0 & 0 & 2/3 & 0 & 0 & 1/3 & 0 & 0 \\ 0 & 0 & 0 & 1 & 0 & 0 & 0 & 0 & 0 \\ 0 & 0 & 0 & 0 & 1 & 0 & 0 & 0 & 0 \\ 0 & 1 & 0 & 0 & 0 & 0 & 0 & 0 & 0 \\ 0 & 0 & 0 & 0 & 0 & 0 & 1 & 0 & 0 \\ 0 & 0 & 0 & 0 & 0 & 1/2 & 0 & 0 & 1/2 \\ 0 & 0 & 1 & 0 & 0 & 0 & 0 & 0 & 0 \\ 0 & 1/3 & 0 & 0 & 0 & 2/3 & 0 & 0 & 0 \\ 0 & 0 & 0 & 0 & 0 & 1 & 0 & 0 & 0 \end{bmatrix}$

d. $\begin{bmatrix} 1/2 & 0 & 0 & 1/3 & 0 & 0 & 0 & 1/6 \\ 0 & 0 & 1 & 0 & 0 & 0 & 0 & 0 \\ 0 & 1 & 0 & 0 & 0 & 0 & 0 & 0 \\ 0 & 0 & 0 & 1/2 & 0 & 1/2 & 0 & 0 \\ 2/3 & 0 & 0 & 0 & 1/3 & 0 & 0 & 0 \\ 0 & 0 & 0 & 0 & 1 & 0 & 0 & 0 \\ 0 & 0 & 0 & 1/4 & 0 & 0 & 3/4 & 0 \\ 0 & 0 & 0 & 0 & 0 & 0 & 0 & 1 \end{bmatrix}$

9.2 Where Do We Get and How Often?

The technique used to find the h_{ij} is one we have been using regularly for some time now—consider what happens at the first step. Starting in E_i, either the first step is to E_j or it is to somewhere else. If the first step is to E_k with $k \neq j$, the probability of reaching E_j eventually is h_{kj}. Thus we have

$$h_{ij} = p_{ij} + \sum_{k \neq j} p_{ik} h_{kj}.$$

Recalling that there are s states, we see there are s^2 different h_{ij} and consequently s^2 equations of the form just given. In other words, we have s^2 linear equations in s^2 unknowns. Fortunately, not every unknown appears in every equation. Only those h_{ij} with one particular j appear in any one equation. Thus we have in fact s systems each of s linear equations in s unknowns. Each system may be treated separately and used to find all h_{ij} for a certain j. There is a slight problem, however: The equations need not have a unique solution; by themselves they do not always determine the h_{ij}. This problem almost solves itself. We know how to find out which h_{ij} are zero. As a matter of common sense, it is reasonable to replace by zero any h_{ij} that we already know to be zero. It can be shown that the equations we have after this is done will always have a unique solution.

Let us carry out the instructions of the last paragraph in an example. It is a good idea for the reader to try to understand why each step we take is correct, but it is also possible just to follow along mechanically. For two reasons, we choose an example where the transition matrix has lots of zeros. In the first place, since we shall be solving systems of linear equations by hand, it is important to keep the computations simple. But also it is important to illustrate the situations that arise when certain states cannot be reached from certain other states. Suppose we have a Markov chain with transition matrix

$$\begin{bmatrix} 1/2 & 1/3 & 1/6 & 0 & 0 \\ 2/5 & 1/5 & 0 & 2/5 & 0 \\ 0 & 0 & 1 & 0 & 0 \\ 0 & 0 & 0 & 2/3 & 1/3 \\ 0 & 0 & 0 & 3/4 & 1/4 \end{bmatrix}.$$

Our goal is to find all the h_{ij}.

We are instructed to treat all h_{ij} with the same last subscript together. Thus we start by finding h_{11}, h_{21}, h_{31}, h_{41}, h_{51}. Since we are discussing the chances of ever reaching E_1, it is not surprising that we treat the numbers in the first column of the transition matrix differently from those in the other columns. Starting in E_1, there is a probability of $1/2$ of reaching E_1 again in one step. There is a probability of $1/3$ of reaching E_2 in the first step and a probability of h_{21} of going on from there to E_1 later. Similarly for $1/6$, E_3, and h_{31}. Thus we have

$$h_{11} = 1/2 + (1/3)h_{21} + (1/6)h_{31}.$$

Analogous reasoning gives us

$$h_{21} = 2/5 + (1/5)h_{21} + \qquad (2/5)h_{41},$$
$$h_{31} = \qquad\qquad\qquad h_{31},$$
$$h_{41} = \qquad\qquad\qquad (2/3)h_{41} + (1/3)h_{51},$$
$$h_{51} = \qquad\qquad\qquad (3/4)h_{41} + (1/4)h_{51}.$$

We have five equations in five unknowns. But the third equation, $h_{31} = h_{31}$, is useless. And each of the last two equations simplifies to $h_{41} = h_{51}$. Thus these five equations by themselves are not enough to determine the h_{ij}. However, either by determining the $\mathcal{R}_i$ or by inspection, we can find out which h_{ij} are zero. In the case at hand, clearly $h_{31} = 0$, since we can never leave E_3. Likewise from E_4 and E_5 only E_4 and E_5 can be reached. Thus, $h_{41} = h_{51} = 0$. Inserting these zeros in our equations we have the new system,

$$h_{11} = 1/2 + (1/3)h_{21},$$
$$h_{21} = 2/5 + (1/5)h_{21}.$$

We can solve these equations easily finding $h_{21} = 1/2$ and then $h_{11} = 2/3$.

Next we find h_{12}, h_{22}, h_{32}, h_{42}, h_{52}. Learning by experience, we write down at once $h_{32} = h_{42} = h_{52} = 0$. Now we write out only the useful equations. These are

$$h_{12} = (1/2)h_{12} + 1/3,$$
$$h_{22} = (2/5)h_{12} + 1/5.$$

Solving, we find $h_{12} = 2/3$ and $h_{22} = 7/15$.

As for those h_{ij} with $j = 3$, the ones that obviously are zero are h_{43} and h_{53}. Our procedure therefore calls for writing the equations,

$$h_{13} = (1/2)h_{13} + (1/3)h_{23} + 1/6,$$
$$h_{23} = (2/5)h_{13} + (1/5)h_{23},$$
$$h_{33} = \qquad\qquad\qquad 1.$$

Solving we have $h_{13} = 1/6$, $h_{23} = 1/2$, $h_{33} = 1$.

We next do h_{ij} with $j = 4$. Clearly $h_{34} = 0$. Our equations thus are

$$h_{14} = (1/2)h_{14} + (1/3)h_{24},$$
$$h_{24} = (2/5)h_{14} + (1/5)h_{24} + 2/5,$$
$$h_{44} = \qquad\qquad 2/3 + (1/3)h_{54},$$
$$h_{54} = \qquad\qquad 3/4 + (1/4)h_{54}.$$

The first two equations yield $h_{14} = 1/2$, $h_{24} = 3/4$. The last two equations yield $h_{44} = 1$, $h_{54} = 1$. (Alternatively, we could have found h_{44} and h_{54} another way. The abstract principle involved is the following: If E_i is recurrent, then either $h_{ij} = 0$ or $h_{ij} = 1$; nothing in between is possible. Because, assuming a start in E_i, there will be infinitely many returns to E_i; if it is at all possible to reach E_j, sooner or later E_j will be reached. In the case at hand, E_4 is recurrent, and thus $h_{44} \neq 0$. Also $h_{54} \neq 0$ since $p_{54} \neq 0$, and E_5 is recurrent. It follows that $h_{44} = h_{54} = 1$.)

Finally we treat the h_{ij} with $j = 5$. Clearly $h_{35} = 0$. We have the equations,

$$h_{15} = (1/2)h_{15} + (1/3)h_{25},$$
$$h_{25} = (2/5)h_{15} + (1/5)h_{25} + (2/5)h_{45},$$
$$h_{45} = \qquad\qquad (2/3)h_{45} + 1/3,$$
$$h_{55} = \qquad\qquad (3/4)h_{45} + 1/4.$$

The last two equations yield $h_{45} = 1$ and $h_{55} = 1$. (The alternative method of the last paragraph may be used here if we prefer.) Then the first two equations give us $h_{15} = 1/2$ and $h_{25} = 3/4$. (That $h_{14} = h_{15}$ and $h_{24} = h_{25}$ is not a coincidence. As already noted, $h_{45} = 1$ and $h_{54} = 1$. Thus, if we reach either of E_4 and E_5, we shall surely reach the other one later. Thus the probability of reaching E_4, from any starting point, is the same as the probability of reaching E_5.)

When E_j is transient, there can be only finitely many visits to E_j over the course of time. Under these circumstances, and only then, it is reasonable to consider the expected number of visits to E_j after a start in E_i. We denote this number by v_{ij}. Thus, v_{ij} is defined only when $h_{jj} < 1$. If $i = j$, the question arises of whether to count the start in E_i as a first visit to E_i. In the end, although not immediately, it turns out to be simpler to say that "after" really means "after." Thus the start in E_i does not count, and v_{ii} is the number of *returns* to E_i after a start there.

Before computing the v_{ij} in general, we consider the case with $i = j$. First note that v_{ii} is defined only when E_i is transient, in other words, only when $h_{ii} < 1$. We start in E_i. Perhaps there is a return visit to E_i; perhaps not. If there is, we repeat the process. We may thus consider a system of Bernoulli trials. A success is leaving E_i for good; a failure is a return to E_i. In standard notation, we have $p = 1 - h_{ii}$ and $q = h_{ii}$. The average number of trials to get one success is $1/p$. Since only the last trial is a success, there are on the average $1/p - 1$ failures before the first success. A failure is a return to E_i. Thus the average number of returns to E_i is $1/p - 1$. In symbols,

$$v_{ii} = \frac{1}{p} - 1.$$

Thus we have,

$$v_{ii} = \frac{1 - p}{p} = \frac{q}{p} = \frac{h_{ii}}{1 - h_{ii}}.$$

Now we consider v_{ij} with $i \neq j$. Assuming a start in E_i, we have h_{ij} as the probability of eventually reaching E_j. If E_j is never reached, the number of visits to E_j is zero. If E_j is reached, we have the first visit plus, by the last paragraph, an average of $h_{jj}/(1 - h_{jj})$ return visits. Thus

$$v_{ij} = (1 - h_{ij}) \cdot 0 + h_{ij}\left(1 + \frac{h_{jj}}{1 - h_{jj}}\right) = h_{ij}\frac{1 - h_{jj} + h_{jj}}{1 - h_{jj}} = \frac{h_{ij}}{1 - h_{jj}}.$$

Now we see that the formula

$$v_{ij} = \frac{h_{ij}}{1 - h_{jj}}$$

holds whenever E_j is transient. We verified it for $i = j$ in the next-to-last paragraph and for $i \neq j$ in the last paragraph.

It is hardly necessary to give an example illustrating the computation of the v_{ij}; after we have found the h_{ij}, we just substitute in a simple formula. Nevertheless, we do return to the example where we found the h_{ij}. Recall v_{ij} is defined only when E_j is transient. In the example, that means for $j = 1$ and $j = 2$. Furthermore, if h_{ij} is zero, then v_{ij} is zero; that conclusion may be obtained either from the definition of v_{ij} or from the formula. Thus, besides noting which v_{ij} are zero, we need only compute

$$v_{11} = \frac{2/3}{1 - 2/3} = 2,$$

$$v_{21} = \frac{1/2}{1 - 2/3} = 3/2,$$

$$v_{12} = \frac{2/3}{1 - 7/15} = 5/4,$$

$$v_{22} = \frac{7/15}{1 - 7/15} = 7/8.$$

Exercises

9. For the transition matrix

$$\begin{bmatrix} 1 & 0 & 0 & 0 \\ 1/2 & 1/4 & 1/4 & 0 \\ 0 & 1/2 & 0 & 1/2 \\ 0 & 0 & 0 & 1 \end{bmatrix},$$

find:

a. h_{21}.

b. h_{22}.

c. All of the other h_{ij}.

10. For the transition matrix of the last problem, find all v_{ij}.

11. For the transition matrix

$$\begin{bmatrix} 1 & 0 & 0 & 0 & 0 \\ 1/4 & 1/4 & 1/2 & 0 & 0 \\ 0 & 1/2 & 0 & 1/2 & 0 \\ 0 & 0 & 1/3 & 1/3 & 1/3 \\ 0 & 0 & 0 & 0 & 1 \end{bmatrix},$$

find

a. h_{21}.

b. h_{22}.

c. all the other h_{ij}.

12. For the transition matrix of the last problem, find all v_{ij}.

13. For the transition matrix

$$\begin{bmatrix} 1 & 0 & 0 & 0 \\ 1/4 & 1/4 & 1/4 & 1/4 \\ 0 & 1/4 & 1/4 & 1/2 \\ 0 & 0 & 0 & 1 \end{bmatrix},$$

find

a. h_{21}.

b. h_{22}.

c. all the other h_{ij}.

14. For the transition matrix of the last problem, find all v_{ij}.

15. For the transition matrix

$$\begin{bmatrix} 1 & 0 & 0 & 0 & 0 \\ 1/2 & 1/2 & 0 & 0 & 0 \\ 0 & 1/4 & 1/4 & 1/4 & 1/4 \\ 0 & 0 & 0 & 1 & 0 \\ 0 & 0 & 0 & 0 & 1 \end{bmatrix},$$

find

a. h_{21}.

b. h_{22}.

c. all the other h_{ij}.

16. For the transition matrix of the last problem, find all v_{ij}.

17. Find all v_{ij} for the transition matrix

$$\begin{bmatrix} 1/2 & 1/2 & 0 & 0 & 0 \\ 1/4 & 3/4 & 0 & 0 & 0 \\ 0 & 0 & 1/3 & 2/3 & 0 \\ 0 & 0 & 1/2 & 1/2 & 0 \\ 0 & 1/3 & 1/3 & 0 & 1/3 \end{bmatrix}.$$

18. Find all v_{ij} for the transition matrix

$$\begin{bmatrix} 1/2 & 1/2 & 0 & 0 \\ 1/2 & 1/2 & 0 & 0 \\ 1/2 & 0 & 1/4 & 1/4 \\ 0 & 1/2 & 0 & 1/2 \end{bmatrix}.$$

19. Find all v_{ij} for the transition matrix

$$\begin{bmatrix} 1/2 & 1/2 & 0 \\ 1/2 & 0 & 1/2 \\ 0 & 0 & 1 \end{bmatrix}.$$

20. Find all v_{ij} for the transition matrix

$$\begin{bmatrix} 1 & 0 & 0 & 0 & 0 \\ 1/3 & 0 & 1/3 & 1/3 & 0 \\ 0 & 0 & 1 & 0 & 0 \\ 0 & 1/3 & 1/3 & 0 & 1/3 \\ 0 & 0 & 0 & 0 & 1 \end{bmatrix}.$$

21. Find all v_{ij} for the transition matrix

$$\begin{bmatrix} 1/3 & 1/3 & 1/3 \\ 0 & 1 & 0 \\ 0 & 1/2 & 1/2 \end{bmatrix}.$$

9.3 How Long Does It Take?

We have already discussed where we go if we start in a certain state. We next discuss how long it takes to get there. Suppose $h_{ij} < 1$ and we start in E_i. There is a possibility that we shall never reach E_j. Thus it makes no sense to speak of the average time needed to reach E_j.

Suppose, on the contrary, $h_{ij} = 1$. Then it can be shown, using the fact that there are only finitely many states in all, that there is a finite average number of steps needed to reach E_j after a start in E_i. We define r_{ij} to be the expected number of steps to reach E_j for the first time after a start in E_i. Note particularly that r_{ii} is the number of steps necessary to be in E_i *again* after a start in E_i. Thus r_{ii} is defined only when $h_{ii} = 1$, that is, when E_i is recurrent.

The method used to evaluate the r_{ij} is quite like the one we used earlier to find the h_{ij}. We fix our attention on a particular r_{ij} first. Since r_{ij} is defined, we have $h_{ij} = 1$. We are finding how long it takes to get from E_i to E_j. Obviously it takes at least one step. If the first step is not to E_j, then, and only then, additional steps are necessary. In more detail, if the first step is to E_k with $k \neq j$, then we need an average of r_{kj} additional steps. Note the r_{kj} must be defined in these circumstances: We know in advance that starting in E_i we must eventually reach E_j. We also know that it is possible to get from E_i to E_k in one step and that $k \neq j$. Thus it must be certain that, having gotten to E_k, we shall go on to E_j. In short, if r_{ij} is defined, $k \neq j$, and $p_{ik} > 0$; then r_{kj} is defined. We repeat what we were saying. To get from E_i to E_j takes at least one step. If the first step is to E_k with $k \neq j$, then r_{kj} additional steps are needed. In *ad hoc* notation we have

$$r_{ij} = 1 + \sum_{}^{*} p_{ik} r_{kj}.$$

By $\sum^{*}$ we mean, "sum over all k for which $k \neq j$ and $p_{ik} \neq 0$." Note that the equations are almost, but not quite, identical to those we used to find the h_{ij}. As with the h_{ij}, those r_{ij} with a particular j are determined together. The equations of the form just given that involve these r_{ij} always constitute a system of linear equations that can be shown to have a unique solution.

We give an example of finding the r_{ij}. Suppose our Markov chain has transition matrix

$$\begin{bmatrix} 1/2 & 0 & 0 & 1/2 \\ 0 & 0 & 1 & 0 \\ 0 & 1/4 & 3/4 & 0 \\ 0 & 1/5 & 0 & 4/5 \end{bmatrix}.$$

The first step is to determine for which i and j there is an r_{ij}. Recall r_{ij} is defined only when $h_{ij} = 1$. It might appear that we therefore need to compute all of the h_{ij} as a first step towards finding the r_{ij}. We are prepared to do just that if necessary. In an example as simple as the one we are working, however, it is clear by inspection which h_{ij} are 1. Since E_1 cannot be reached from any other state, we have $h_{21} = h_{31} = h_{41} = 0$. From E_1 there is a chance of going to E_4, from which return to E_1 is impossible; thus $h_{11} < 1$. We conclude r_{ij} is not defined for any i if $j = 1$. Now consider $j = 4$. It is clear that, from E_1, sooner or later we will go to E_4. Thus $h_{14} = 1$ and r_{14} is defined. Since E_4 cannot be reached at all from either E_2 or E_3, we see that both r_{24} and r_{34} are meaningless. Wherever we start, it is clear that eventually we shall be wandering between E_2 and E_3, reaching each of them from time to time; thus, $h_{ij} = 1$ for $j = 2, 3$. Therefore, r_{ij} is defined provided $j = 2$ or 3. To summarize, r_{12}, r_{13}, r_{14}, r_{22}, r_{23}, r_{32}, r_{33}, r_{42}, r_{43}, are the only r_{ij} that make sense.

We next evaluate the r_{ij}. We begin with $j = 2$. In doing that, we treat the second column of the transition matrix differently from the other columns. Since $p_{12} = 0$, to get from E_1 to E_2 we make a first step and then a number of additional steps, how many additional steps depending on where we went with the first step. Either by looking back at the last paragraph, or by thinking it out, we see

$$r_{12} = 1 + (1/2)r_{12} + \qquad\qquad (1/2)r_{42}.$$
$$r_{22} = 1 + \qquad\qquad r_{32},$$
$$r_{32} = 1 + \qquad (3/4)r_{32},$$
$$r_{42} = 1 + \qquad\qquad (4/5)r_{42}.$$

In writing the last two equations, we used the fact that if the first step is to E_2, no additional steps are necessary to reach E_2. We find in turn $r_{32} = 4$, $r_{42} = 5$, $r_{22} = 5$, $r_{12} = 7$. The next set of equations, for $j = 3$, is

$$r_{13} = 1 + (1/2)r_{13} + \qquad\qquad (1/2)r_{43},$$
$$r_{23} = 1,$$
$$r_{33} = 1 + \qquad (1/4)r_{23},$$
$$r_{43} = 1 + \qquad (1/5)r_{23} + (4/5)r_{43}.$$

Solving we find $r_{23} = 1$, $r_{43} = 6$, $r_{13} = 8$, $r_{33} = 5/4$. The last set of equations, for $j = 4$, consists of just one equation,

$$r_{14} = 1 + (1/2)r_{14}.$$

The solution is $r_{14} = 2$.

We can apply the method just described to the following problem: How many Bernoulli trials are necessary, on the average, to get c consecutive successes? It takes a little care to put this problem in terms of a Markov chain. The key point can be described by an example. Suppose $c = 7$, that is, we are trying to get seven successes in a row. Assume that at some time the last three trials were all successes. It makes no difference whether these three trials were immediately preceded by a failure or they were the first three trials. We do want to distinguish those two possibilities from the situation where the three successes were immediately preceded by another success.

Now we're ready to define a Markov chain. In this one example, it is convenient to number the states from 0 to c, contrary to our usual practice of numbering the states from 1 to s. For each $i = 0, \ldots, c - 1$, we may define the state E_i as follows: To be in E_i is to just have completed exactly i successes in a row. Thus we start in E_0 and return to E_0 after each failure. Put the other way round, to be in E_i means we have $c - i$ successes to go to complete the c consecutive successes we are seeking. It makes no difference what happens after we have obtained the c consecutive successes. We may as well introduce an absorbing state for the situation "game over"; we call this state E_c. Thus we have a Markov chain with $c + 1$ states, and we seek r_{0c}, the average number of steps necessary to reach E_c after a start in E_0.

Towards finding r_{0c}, the first step is to find the transition matrix. Suppose we are in E_i with $0 \leq i \leq c - 1$. If the next trial results in success, we move to E_{i+1}, that is, $p_{i,i+1} = p$. (Of course, p and q have their usual meanings for Bernoulli trials.) If the next trail results in failure, we go back to E_0, that is, $p_{i0} = q$. Necessarily then, for $0 \leq i \leq c - 1$, $p_{ij} = 0$ if $j \neq 0$ and $j \neq i + 1$. Clearly, $p_{cj} = 0$ unless $j = c$, but $p_{cc} = 1$. We write out the equations that we suggested above as a means of finding the r_{ij}; here we want $j = c$, of course. Using the values for p_{ij} just given, we have

$$r_{ic} = 1 + pr_{i+1,c} + qr_{0c},$$

provided $0 \leq i \leq c - 2$. For $i = c - 1$, we have

$$r_{c-1,c} = 1 + qr_{0c}.$$

For brevity, we set $x_i = r_{c-i,c}$ for $i = 1, \ldots, c$. Thus we seek to find x_c by using the system of equations:

$$
\begin{aligned}
x_1 &= 1 & &+ qx_c, \\
x_2 &= 1 + px_1 & &+ qx_c, \\
x_3 &= 1 + px_2 & &+ qx_c, \\
&\;\;\vdots \\
x_c &= 1 + px_{c-1} + qx_c.
\end{aligned}
\qquad (*)
$$

We can simplify matters by subtracting each equation from its successor. We have then

$$
\begin{aligned}
x_2 - x_1 &= & & px_1, \\
x_3 - x_2 &= p(x_2 - x_1) &=& \quad p^2 x_1, \\
x_4 - x_3 &= p(x_3 - x_2) &=& \quad p^3 x_1, \\
&\;\;\vdots \\
x_c - x_{c-1} &= p(x_{c-1} - x_{c-2}) = p^{c-1} x_1.
\end{aligned}
$$

Adding we have

$$
\begin{aligned}
x_c - x_1 &= x_1 (p + p^2 + \cdots + p^{c-1}), \\
x_c &= x_1 (1 + p + p^2 + \cdots + p^{c-1}).
\end{aligned}
$$

Using the first equation of $(*)$ again, we have

$$
\begin{aligned}
x_c &= (1 + qx_c)(1 + p + \cdots + p^{c-1}), \\
qx_c &= (1 + qx_c)(1 - p)(1 + p + \cdots + p^{c-1}), \\
&= (1 + qx_c)(1 - p^c). \\
&= 1 + qx_c - p^c - p^c qx_c.
\end{aligned}
$$

Hence

$$x_c = \frac{1 - p^c}{p^c q}.$$

That answers our question; $(1 - p^c)/(p^c q)$ is the average number of Bernoulli trials necessary to get c successes in a row.

We started this chapter by announcing we were going to generalize the ideas of the last chapter. Recall that in the last chapter one question discussed was how long it takes for someone, either Peter or Paul, to go broke. We answered that question even though we didn't know which player would go broke. In our present language, we were finding, for certain Markov chains, how long it takes to before an absorbing state is reached. In other words, how many of the steps taken go to transient states?

We may as well generalize by considering an arbitrary Markov chain. As we have already noted, there will come a time after which only recurrent states are visited. How long does it take for this situation to occur? We already have at hand almost all the machinery needed to answer this question.

Our first chore, however, remains to be described. This chore consists of replacing the given Markov chain with a new chain. The transient states are left unchanged. We replace all the recurrent states by a single state, E^*. Since the new states are described in terms of the old, the transition probabilities are just the already-known probabilities of going from one state to another. To spell this out, the probability of going from any transient state to another remains the same. The new state E^* is to be absorbing, since it is impossible to go from a recurrent state to a transient state. The probability of going from a transient state E_i to E^* is the sum of the probabilities of going from E_i to the various recurrent states of the old chain. At the risk of undue repetition, let us state what is to be done mechanically. We assume the new state E^* is to be considered the last state corresponding to the bottom row and right-hand column of the matrix. The first step is to strike out from the given matrix the rows and columns corresponding to the recurrent states. Then a new column is inserted at the right; the entries in this new column are determined by the fact that the total of each row must be one. Finally a new row, consisting of 0s followed by a single 1, is added at the bottom. Now we are ready to use the method developed earlier in this section.

At this point, we have completely described a method of finding the answer to the question raised two paragraphs back: Given a Markov chain, for each transient state, assuming a start in that state, what is the average number of steps needed to reach a recurrent state? First we modify the transition matrix in the manner just described. That r_{ij} that corresponds to going from the given transient state to the new state E^* is the answer to our question.

Let us consider an example. Suppose a Markov chain has transition matrix

$$\begin{bmatrix} 1 & 0 & 0 & 0 & 0 & 0 \\ 1/2 & 0 & 1/3 & 0 & 1/6 & 0 \\ 0 & 0 & 1/2 & 1/2 & 0 & 0 \\ 0 & 0 & 1/4 & 3/4 & 0 & 0 \\ 0 & 1/7 & 0 & 0 & 2/7 & 4/7 \\ 0 & 0 & 0 & 0 & 0 & 1 \end{bmatrix}.$$

E_1 and E_6 are absorbing, and hence recurrent. E_3 and E_4 are clearly also recurrent. Now we see that E_2 and E_5 are transient. Accordingly we delete the first, third, fourth, and sixth rows and the corresponding columns. Now we have the matrix

$$\begin{bmatrix} 0 & 1/6 \\ 1/7 & 2/7 \end{bmatrix}.$$

This is not a transition matrix, but we make it into one by inserting an extra row and column:

$$\begin{bmatrix} 0 & 1/6 & 5/6 \\ 1/7 & 2/7 & 4/7 \\ 0 & 0 & 1 \end{bmatrix}.$$

For this last matrix we have

$$r_{13} = 1 + (1/6)r_{23},$$
$$r_{23} = 1 + (1/7)r_{13} + (2/7)r_{23}.$$

From this we find $r_{23} = 48/29$ and $r_{13} = 37/29$. In other words, returning to the original Markov chain, it takes an average of $48/29$ steps to reach a recurrent state if we start in E_5, and an average of $37/29$ steps if we start in E_2.

Exercises

22. Find all r_{ij} for the transition matrix

$$\begin{bmatrix} 1/2 & 1/2 & 0 \\ 1/2 & 0 & 1/2 \\ 0 & 0 & 1 \end{bmatrix}.$$

23. Find all r_{ij} for the transition matrix

$$\begin{bmatrix} 0 & 1/2 & 1/2 \\ 1 & 0 & 0 \\ 0 & 1 & 0 \end{bmatrix}.$$

24. Find all r_{ij} for the transition matrix

$$\begin{bmatrix} 1/3 & 1/3 & 1/3 \\ 0 & 1 & 0 \\ 0 & 1/2 & 1/2 \end{bmatrix}.$$

25. Find all r_{ij} for the transition matrix

$$\begin{bmatrix} 1/2 & 1/2 & 0 & 0 \\ 1/2 & 1/2 & 0 & 0 \\ 1/2 & 0 & 1/4 & 1/4 \\ 0 & 1/2 & 1/4 & 1/2 \end{bmatrix}.$$

26. Find all r_{ij} for the transition matrix

$$\begin{bmatrix} 1/2 & 0 & 1/2 \\ 0 & 1/2 & 1/2 \\ 0 & 1/3 & 2/3 \end{bmatrix}.$$

27. Find all r_{ij} for the transition matrix

$$\begin{bmatrix} 1 & 0 & 0 & 0 & 0 \\ 0 & 1 & 0 & 0 & 0 \\ 0 & 0 & 1/2 & 1/2 & 0 \\ 0 & 0 & 1/2 & 1/2 & 0 \\ 0 & 1/3 & 0 & 1/3 & 1/3 \end{bmatrix}$$

28. Find all r_{ij} for the transition matrix

$$\begin{bmatrix} 1/2 & 1/2 \\ 1/3 & 2/3 \end{bmatrix}$$

29. Find all r_{ij} for the transition matrix

$$\begin{bmatrix} 1/2 & 1/2 & 0 \\ 3/4 & 0 & 1/4 \\ 0 & 1/4 & 3/4 \end{bmatrix}$$

30. Show that the formulas of Chapters Four and Nine agree as to the number of Bernoulli trials needed to get one success.

31. a. Assuming a coin to be tossed once a second, how long would it take, on the average, of $c = 5, 10, 15, 20, 25, 30, 40, 50$ to get c heads in a row?

   b. Assuming a die to be thrown once a second, how long would it take, on the average, for each of $c = 5, 10, 15, 20$ to get c "sixes" in a row?

   c. Assuming a die to be thrown once a second, how long would it take, on the average, for each of $c = 5, 10, 25, 50, 75, 100, 125, 150, 175, 200$ to get c consecutive throws none of which is a "six"?

32. Letters are chosen at random, with replacement, from the word

   BANANA

   until each of the three different letters B, A, and N has been drawn at least once. Find the expected number of times a letter is drawn. (Hint: Assign a state to each of "nothing," B, A, N, BA, BN, AN, BAN.)

33. On the average, how many times must one throw a die to obtain a "one," a "two," and a "three," in that order, on consecutive throws? Suggestion: Let E_4 be "mission accomplished." Let E_3 be "one" and "two" just thrown, "three" still needed. Let E_2 be "one" just thrown, "two" and "three" still needed. Let E_1 be nothing useful yet done. Be careful; note for example: If you are in E_3, needing only a "three," and you throw a "one," then you go to E_2, not to E_1.

34. On the average, how many times must one throw a die to obtain a "one," a "two," and another "one," in that order, on consecutive throws? Suggestion: Modify your solution to the last problem.

35. On the average, how many times must one throw a die to obtain a "one," another "one," and a "two," in that order, on consecutive throws?

36. On the average, how many times must one throw a die to obtain a "one," a "two," and another "two," in that order, on consecutive throws?

37. On the average, how many times must one throw a die to obtain three consecutive throws that fall alike?

38. On the average, how many times must one throw a die to obtain three consecutive throws no two of which fall alike?

39. A coin is tossed repeatedly until it falls heads–tails–heads, in that order, on three consecutive tosses. Find the expected number of tosses necessary.

40. A coin is tossed repeatedly until it falls tails–tails–heads, in that order, on three consecutive tosses. Find the expected number of tosses necessary.

41. A coin is tossed repeatedly until it falls heads–tails–tails, in that order, on three consecutive tosses. Find the expected number of tosses necessary.

42. On the average, how many times must one toss a coin to obtain three consecutive tosses that fall alike?

43. On the average, how many times must one toss a coin to obtain three consecutive tosses that do not fall all alike?

44. On the average, how many times must one throw a die to get six consecutive tosses that fall alike?

**45. A die is thrown repeatedly until it falls "one," "two," "three," "four," "five," "six," in that order, on consecutive throws. Find the expected number of throws made.

**46. A die is thrown repeatedly until six consecutive throws show six different numbers. Find the expected number of throws made.

47. In the situation in Exercise 3, if the customer buys Brand A now, how long will it take before the customer buys Brand D?

48. In the situation of Exercise 1, if a painting by Renoir is on display now, how long will it be, on the average, before a painting by Cézanne is on display?

49. Do the last problem for the situation of Exercise 2.

**50. In the situation of Exercise 4, for each possible starting point, how many of the rooms does Mr. Nuff pass through, on the average, before he leaves the museum?

**51. In the situation of Exercise 7, if an account is now current, how long will it be, on the average, before the account is closed?

52. For each state, assuming a start in that state, how long does it take, on the average, to reach a recurrent state for the transition matrix

$$\begin{bmatrix} 1 & 0 & 0 & 0 \\ 1/4 & 1/4 & 1/4 & 1/4 \\ 0 & 1/4 & 1/4 & 1/2 \\ 0 & 0 & 0 & 1 \end{bmatrix}.$$

53. For each state, assuming a start in that state, how long does it take, on the average, to reach a recurrent state for the transition matrix

$$\begin{bmatrix} 1 & 0 & 0 & 0 \\ 1/2 & 1/4 & 1/4 & 0 \\ 0 & 1/2 & 0 & 1/2 \\ 0 & 0 & 0 & 1 \end{bmatrix}.$$

54. For each state, assuming a start in that state, how long does it take, on the average, to reach a recurrent state for the transition matrix

$$\begin{bmatrix} 1 & 0 & 0 & 0 & 0 \\ 1/5 & 4/5 & 0 & 0 & 0 \\ 0 & 1/3 & 1/3 & 1/3 & 0 \\ 0 & 1/4 & 0 & 1/2 & 1/4 \\ 0 & 0 & 0 & 0 & 1 \end{bmatrix}.$$

55. For each state, assuming a start in that state, how long does it take, on the average, to reach a recurrent state for the transi-

matrix

$$\begin{bmatrix} 1/2 & 1/2 & 0 & 0 \\ 1/2 & 1/2 & 0 & 0 \\ 1/2 & 0 & 1/4 & 1/4 \\ 0 & 0 & 1/2 & 1/2 \end{bmatrix}.$$

56. For each state, assuming a start in that state, how long does it take, on the average, to reach a recurrent state for the transition matrix

$$\begin{bmatrix} 1 & 0 & 0 & 0 & 0 & 0 \\ 0 & 0 & 1/2 & 1/2 & 0 & 0 \\ 0 & 1/3 & 0 & 2/3 & 0 & 0 \\ 0 & 1/4 & 1/2 & 1/4 & 0 & 0 \\ 1/4 & 0 & 0 & 0 & 1/2 & 1/4 \\ 0 & 1/2 & 0 & 0 & 1/2 & 0 \end{bmatrix}.$$

9.4   What Happens in the Long Run?

Before getting to the main subject matter of this section, which concerns what happens in a Markov chain after a recurrent state is reached, we introduce some notation. There is an obvious generalization of the numbers p_{ij} introduced earlier. Recall p_{ij} is the conditional probability, given being in E_i at some stage, of being in E_j one stage later. We now let $p_{ij}(n)$ be the conditional probability, given being in E_i at some stage, of being in E_j just n stages later. Thus, $p_{ij}(1)$ means the same as p_{ij}. It is not surprising that the $p_{ij}(n)$ can be computed easily from the p_{ij}, in other words, from the transition matrix. We start with the $p_{ij}(2)$; in fact, think of just one $p_{ij}(2)$. We seek the probability, assuming we start in E_i, of being in E_j after two steps. At the first step, we do go somewhere. We find $p_{ij}(2)$ by summing over all k the probabilities of going from E_i to E_k in one step and then going on to E_j at the next step. In symbols,

$$p_{ij}(2) = \sum_k p_{ik}p_{kj}.$$

These formulas may be used to find all the $p_{ij}(2)$. There are two corresponding formulas for the $p_{ij}(3)$. To get from E_i to E_j in three steps, we may go from E_i to E_k in two steps and then on to E_j in one additional step. Summing over all k, we have

$$p_{ij}(3) = \sum_k p_{ik}(2)p_{kj}.$$

Alternatively, we can consider going from E_i to E_k in a single step and then on to E_j in two additional steps. Thus,

$$p_{ij}(3) = \sum_k p_{ik}p_{kj}(2).$$

To find $p_{ij}(4)$, we may use any one of three different formulas. They are

$$p_{ij}(4) = \sum_k p_{ik}(3)p_{kj},$$

$$p_{ij}(4) = \sum_k p_{ik}(2)p_{kj}(2),$$

$$p_{ij}(4) = \sum_k p_{ik}p_{kj}(3).$$

There is no point in trying to describe all formulas along these lines; they are really quite obvious. We simply record one fact that we shall need later; for all n, i, and j,

$$p_{ij}(n+1) = \sum_k p_{ik}(n)p_{kj}.$$

[This paragraph is intended for those readers who happen to know the definition of matrix multiplication. These readers will recognize the similarity between that definition and the formulas above. It is evident that, for each n, the matrix that has $p_{ij}(n)$ as the entry in the ith row, jth column is simply T^n, the nth power of the transition matrix T.]

Given a transition matrix, the computation of any particular $p_{ij}(n)$ is easy, though perhaps tedious. It is also an easy task for a computer. We shall not give exercises that involve making this computation since our only goal in introducing the notation was to use it in discussing theory.

Let us fix our attention on all $p_{ij}(n)$ with a particular i and n. In other words, we assume a start in E_i and consider what happens n steps later. The numbers

$$p_{i1}(n), p_{i2}(n), \ldots, p_{is}(n)$$

are the probabilities of being in the various states exactly n steps after a start in E_i. Thus, of course, each of these numbers is nonnegative, and the numbers total 1. We generalize. Any list of s nonnegative numbers that total 1 will be called a *probability vector*. (The word vector is used here simply to indicate that we have a list of numbers.) To restate the definition, $(a_1, a_2, \ldots, a_s)$ is a probability vector if

$$0 \leq a_j \quad \text{for all } j = 1, \ldots, s; \quad \text{and} \sum_{j=1}^{s} a_j = 1.$$

Suppose the probability vector $(a_1, \ldots, a_s)$ gives the probabilities of being in the various states at some time. We mean, of course, that a_j is the probability of being in E_j at that time. For convenience, call the time in question "now." What is the probability b_j of being in E_j exactly n steps from now? The conditional probability, given being in E_i now, of being in E_j after n steps is $p_{ij}(n)$. Since we must be somewhere now,

$$b_j = \sum_i a_i p_{ij}(n).$$

In the special case $n = 1$, we see that b_j is the probability of being in E_j one step from now, and

$$b_j = \sum_i a_i p_{ij}.$$

We next give a definition whose importance will become obvious in the discussion that follows it. We call a probability vector $(m_1, \ldots, m_s)$ *fixed* if the assumption that the numbers $m_1, \ldots, m_s$ are the probabilities of being in the respective states at some time leads to the conclusion that these same numbers are also the probabilities of being in those states one step later. In other words, a probability vector $(m_1, \ldots, m_s)$ is called fixed if

$$m_j = \sum_i m_i p_{ij} \quad \text{for all } j.$$

Of course, if one step does not change the probabilities, a second step will not change them either. Thus, if $(m_1, \ldots, m_s)$ is a fixed probability vector that gives the probability of being in each of the states at a certain time, it also gives these probabilities at every later time. In symbols,

$$m_j = \sum_i m_i p_{ij}(n)$$

for all n.

At present, we do not know which Markov chains have a unique fixed probability vector, but, given such a chain, finding the fixed probability vector is easy. We may as well discuss how to do that now, even though we do not yet know the significance of such vectors. A fixed probability vector $(m_1, \ldots, m_s)$ is simply a solution of the linear equations

$$m_j = \sum_i m_i p_{ij},$$
$$\sum_i m_i = 1.$$

Thus we need merely solve these linear equations in the unknowns $m_1, \ldots, m_s$. No harm is done by the fact that we have s unknowns and $s + 1$ equations. (Working by hand, we just ignore some of the information. If we seek to use a computer program that allows for only as many equations as unknowns, we simply omit any one equation except the last. It is easy to see that the omitted equation will automatically be satisfied by any solution of the others.) Thus, assuming the equations have a unique solution, we may find this solution in a routine way.

We give an example. Consider the transition matrix

$$\begin{bmatrix} 2/3 & 1/3 & 0 \\ 1/4 & 1/2 & 1/4 \\ 0 & 4/5 & 1/5 \end{bmatrix}.$$

We shall show that there is just one fixed probability vector and find this vector. Suppose x, y, and z are the probabilities of being in E_1, E_2, and E_3 at a particular time. Let us compute the probability of being in E_1 one stage later. We can get to E_1 directly from E_1; the probability we are in E_1 and then stay in E_1 is $x(2/3)$. Likewise, the probability of being in E_2 and then going to E_1 in a single step is $y(1/4)$. We cannot get from E_3 to E_1 in a single step. Thus the probability of being in E_1 one step after the "particular time" is

$$(2/3)x + (1/4)y.$$

If (x, y, z) is to be a fixed probability vector, we must have

$$x = (2/3)x + (1/4)y.$$

Likewise, we need

$$y = (1/3)x + (1/2)y + (4/5)z,$$
$$z = \qquad (1/4)y + (1/5)z.$$

These last three equations clearly do not determine (x, y, z); $x = y = z = 0$ satisfies all of them but is not the solution we seek. We want x, y, and z with

$$x + y + z = 1.$$

Now we have four equations. We discard the most complicated, that is, the second, of them and solve the others. If we feel cautious, we can check whether the discarded equation is satisfied by our answer, once we find our answer. From $x = (2/3)x + (1/4)y$ we conclude $x = (3/4)y$. From $z = (1/4)y + (1/5)z$, we conclude $z = (5/16)y$. Thus

$$(3/4)y + y + (5/16)y = 1.$$

Thus $y = 16/33$, and hence $x = 4/11$ and $z = 5/33$. Our fixed probability vector is thus $(4/11, 16/33, 5/33)$. We now return to theory and discuss, among other things, the significance of a fixed probability vector.

We are ready now to turn to the main subject of this section, namely, what happens in a Markov chain after we reach a recurrent state? We have already noted that sooner or later we definitely will reach some recurrent state. If this state is absorbing, we just stay there, and nothing worth mentioning occurs further. However, if we reach a recurrent state that is not absorbing, we still have something to study. Of course, one possibility is to start in a recurrent state. As we already know, starting in a certain state is entirely equivalent to reaching this state later, as far as what happens next. Therefore, the discussion that follows does not distinguish between starting in a recurrent state and reaching that state after a while. Our detailed work will all be based on the assumption that a particular recurrent state, which we shall designate E_i, is reached. Since, for some chains and starting points, we do not know in advance which recurrent states will be reached, we may have to apply the discussion that follows to more than one choice of E_i. Now we consider what happens afterwards if we do reach a certain recurrent state E_i.

Let a particular recurrent state E_i be chosen. There typically will be states that cannot be reached from E_i. Since we are concerned only with what happens after being in E_i, we shall soon discard those states; they are of no further interest to us. But there is something we should discuss first. Recall that $\mathcal{R}_i$ is the set of states that can be reached from E_i. Since E_i is recurrent, we can return from any state in $\mathcal{R}_i$ to E_i. It follows we can get from any state in $\mathcal{R}_i$ to any state in $\mathcal{R}_i$. On the other hand, it follows from the definition of $\mathcal{R}_i$ that one cannot get from a state in $\mathcal{R}_i$ to a state that is not in $\mathcal{R}_i$. Now we are ready to discard all states outside of $\mathcal{R}_i$. In mechanical terms, we simply delete from the transition matrix the rows and columns corresponding to all states that cannot be reached from E_i. We then have the transition matrix for a new Markov chain. This new chain has the property that it is possible to get from any state to any state. Studying this new chain is equivalent to discussing what happens in the original chain after we reach E_i.

Until further notice, we shall confine our attention to the new Markov chain introduced in the last paragraph. To say the same thing another way, for the time being we confine our attention to chains in which we can get from any state to any state. Clearly in such a chain all states are recurrent. Over the course of time, wherever we start, we shall pay infinitely many visits to every state. What fraction of the time we spend, on the average, in each state is an obvious question to raise.

The first job to do is to make it clear exactly what we mean by "fraction of the time." We begin doing that by choosing a positive integer N and fixing our attention on the first N steps. Later we can get an overall view by considering the limit as N tends to infinity.

As we just said, let N be a fixed positive integer. Consider certain states E_i and E_j. We assume a start in E_i. We let the random variable X be the number of steps, among the first N steps, that go to E_j. We evaluate X in the usual way. Let random variables $X_1, \ldots, X_N$ be defined as follows:

$$X_n = \begin{cases} 1 & \text{if the } n\text{th step is to } E_j, \\ 0 & \text{otherwise.} \end{cases}$$

Then $\mathbf{E}(X_n) = p_{ij}(n)$, and hence $\mathbf{E}(X) = \mathbf{E}(X_1 + \cdots + X_N) = \mathbf{E}(X_1) + \cdots + \mathbf{E}(X_N) = p_{ij}(1) + \cdots + p_{ij}(N)$. The quantity

$$[p_{ij}(1) + \cdots + p_{ij}(N)]/N$$

is the fraction of the first N steps that go to E_j assuming a start in E_i.

Before letting N tend to infinity, we find another expression for the quantity of the last paragraph. We continue to use the fixed number N and the fixed states E_i and E_j. Now we introduce some numbers of little importance in themselves; we obtain the formula we seek by evaluating these numbers two ways and equating the results. We note that, starting in E_i, over the course of time we will make infinitely many visits to E_j. Thus we may let a_n be the average number of steps until that visit to E_j that occurs first when whatever happens in the first n stages is disregarded. In other words, a_n is the expected number of steps to that first visit to E_j that occurs after $n+1$ or more steps.

We can write down a value for a_n immediately. If $n = 0$, we have $a_n = r_{ij}$. For $n \geq 1$, a_n involves a visit to E_j that occurs after more than n steps. If the first n steps take us to E_k, it will take r_{kj} more steps to reach E_j; thus we have

$$a_n = n + p_{i1}(n)r_{1j} + \cdots + p_{is}(n)r_{sj}.$$

Now we find $a_n - a_{n-1}$ directly, without considering the values of a_n and a_{n-1} separately. Each of a_n and a_{n-1} is the number of steps until the first visit to E_j that occurs after a certain number of steps—after $n + 1$ steps in the first case and after n steps in the second case. $a_n - a_{n-1}$ is the average number of steps between these visits. The same visit is involved in both cases unless it happens that the nth step is to E_j. In this last situation, for a_n we must wait an additional r_{jj} steps to get back to E_j, but no wait is necessary for a_{n-1}. Thus we have

$$a_n - a_{n-1} = p_{ij}(n)r_{jj} + [1 - p_{ij}(n)] \cdot 0 = p_{ij}(n)r_{jj}.$$

We may write

$$a_N = (a_N - a_{N-1}) + (a_{N-1} - a_{N-2}) + \cdots + (a_1 - a_0) + a_0.$$

It follows

$$a_N = p_{ij}(N)r_{\bar{j}j} + p_{ij}(N-1)r_{\bar{j}j} + \cdots + p_{ij}(1)r_{\bar{j}j} + r_{ij}$$
$$= r_{\bar{j}j}[p_{ij}(N) + p_{ij}(N-1) + \cdots + p_{ij}(1)] + r_{ij}.$$

As noted above, we also have

$$a_N = N + p_{i1}(N)r_{1j} + \cdots + p_{is}(N)r_{sj}.$$

Combining the last two equations, we have

$$r_{\bar{j}j}[p_{ij}(N) + p_{ij}(N-1) + \cdots + p_{ij}(1)] + r_{ij} = N + p_{i1}(N)r_{1j} + \cdots + p_{is}(N)r_{sj}.$$

From this we find

$$[p_{ij}(N) + p_{ij}(N-1) + \cdots + p_{ij}(1)]/N$$
$$= \frac{1}{r_{\bar{j}j}}\left[-\frac{r_{ij}}{N} + 1 + \frac{p_{i1}(N)r_{1j} + \cdots + p_{is}(N)r_{sj}}{N}\right].$$

Now we are almost ready to let N tend to infinity. All we have to do first is to note that

$$0 \leq p_{i1}(N)r_{1j} + \cdots + p_{is}(N)r_{sj} \leq r_{1j} + \cdots + r_{sj},$$

which does not depend on N. Thus

$$\lim_{N\to\infty}[p_{i1}(N)r_{1j} + \cdots + p_{is}\mu(N)r_{sj}]/N = 0.$$

Hence, we have

$$[p_{ij}(N) + p_{ij}(N-1) + \cdots + p_{ij}(1)]/N = 1/r_{\bar{j}j}.$$

The limit just evaluated may reasonably be regarded as the fraction of the time spent in E_j after a start in E_i. An important fact stares us in the face: It does not matter where we start; the fraction of the time spent in E_j will be the same for any starting point.

We already have one way to evaluate all the r_{ij}, and hence to find $1/r_{\bar{j}j}$ for all j. After some additional discussion, we shall find that there is another method to find all $1/r_{\bar{j}j}$ and that this other method is a more efficient route to that goal.

We next study the properties of the numbers $1/r_{11}, 1/r_{22}, \ldots, 1/r_{ss}$. Clearly each of these numbers is positive. Towards finding out more, we introduce a temporary abbreviation $x_j(N)$ by letting

$$x_j(N) = [p_{ij}(1) + p_{ij}(2) + \cdots + p_{ij}(N)]/N.$$

We just showed

$$\lim x_j(N) = 1/r_{jj}.$$

We have

$$\sum_j x_j(N) = \sum_j [p_{ij}(1) + \cdots + p_{ij}(N)]/N$$

$$= \sum_j p_{ij}(1)/N + \cdots + \sum_j p_{ij}(N)/N$$

$$= 1/N + \cdots + 1/N = 1.$$

Thus we conclude

$$\sum_j 1/r_{jj} = \sum_j \lim x_j(N) = \lim \sum_j x_j(N) = \lim 1 = 1.$$

In short, $(1/r_{11}, \ldots, 1/r_{ss})$ is a probability vector. Next we consider

$$\sum_k x_k(N)p_{kj} = \sum_k p_{kj}[p_{ik}(1) + \cdots + p_{ik}(N)]/N$$

$$= \sum_k p_{ik}(1)p_{kj}/N + \cdots + \sum_k p_{ik}(N)p_{kj}/N$$

$$= [p_{ij}(2) + \cdots + p_{ij}(N+1)]/N$$

$$= [p_{ij}(1) + \cdots + p_{ij}(N)]/N + [p_{ij}(N+1) - p_{ij}(1)]/N.$$

Since $|p_{ij}(N+1) - p_{ij}(1)| \le 1$,

$$\lim[p_{ij}(N+1) - p_{ij}(1)]/N = 0.$$

Thus

$$\lim \sum_k x_k(N)p_{kj} = 1/r_{jj}.$$

Now we have

$$\sum_k (1/r_{kk})p_{kj} = \sum_k [\lim x_k(N)]p_{kj} = \lim \sum_k x_k(N)p_{kj} = 1/r_{jj}.$$

By definition then, $(1/r_{11}, \ldots, 1/r_{ss})$ is a fixed probability vector.

In the last paragraph we showed there is at least one fixed probability vector, namely, $(1/r_{11}, \ldots, 1/r_{ss})$. Now we show that there no other fixed probability vectors. Consider any fixed probability vector $(m_1, \ldots, m_s)$. Then, as noted above, we have

$$m_j = \sum_k m_k p_{kj}(n) \qquad \text{for all } n.$$

From this it follows that

$$m_j = \left[\sum_k m_k p_{kj}(1) + \sum_k m_k p_{kj}(2) + \cdots + \sum_k m_k p_{kj}(N) \right] /N$$

$$= \sum_k m_k [p_{kj}(1) + p_{kj}(2) + \cdots + p_{kj}(N)]/N \qquad \text{for all } N.$$

Taking the limit, we find

$$m_j = \lim m_j = \lim \sum_k m_k [p_{kj}(1) + p_{kj}(2) + \cdots + p_{kj}(N)]/N$$

$$= \sum_k m_k \lim [p_{kj}(1) + p_{kj}(2) + \cdots + p_{kj}(N)]/N$$

$$= \sum_k m_k (1/r_{jj}) = (1/r_{jj}) \sum_k m_k = 1/r_{jj}.$$

Thus the arbitrary fixed probability vector $(m_1, \ldots, m_s)$ is identical to $(1/r_{11}, \ldots, 1/r_{ss})$. It follows that $(1/r_{11}, \ldots, 1/r_{ss})$ is the only fixed probability vector. Recall we are working under the hypothesis that one can get from any state to any state. The fact that, under this hypothesis, there is a unique fixed probability vector is important in itself.

It is now clear how to find the values of the $1/r_{jj}$ efficiently. All we need do is find a fixed probability vector. This vector will necessarily be $(1/r_{11}, \ldots, 1/r_{ss})$. An example showing how to do this will be found above.

At this point a warning may be in order. We illustrate what we were talking about with an example. Consider the Markov chain with transition matrix:

$$\begin{bmatrix} 0 & 1 \\ 1 & 0 \end{bmatrix}.$$

By direct computation we find that $(1/2, 1/2)$ is the only fixed probability vector. It follows, and it was obvious from the matrix, that wherever we start we spend half the time in each state. But it does not follow that, for example,

$$p_{12}(3456789) = 1/2.$$

If we start in E_1, the probability of being in E_2 exactly 3,456,789 steps later, or any odd number of steps later, is 1. After an even number of steps we would be in E_1. We simply alternate between the two states in a completely predetermined way; chance plays no role. More complicated examples, which do involve chance, can also be given; see Exercises 60–63. While it is often the case that when and where we start makes very little difference after a while, this need not happen. On the other hand, we take nothing back. In the example just given, there is a unique fixed probability vector, and this vector does give, for both possible starting points, the fraction of the time to be spent in each state.

We now return to a general Markov chain. We discuss briefly the question of how many fixed probability vectors exist. First we recall that a transient state can be reached only finitely many times. It follows that every fixed probability must assign the probability zero to each transient state. Suppose that every recurrent state can be reached from each recurrent state. It is clear from the discussion above that assigning zero to each transient state and $1/r_{jj}$ to each recurrent state E_j gives a fixed probability vector in this case. It is also clear that this vector is the only fixed probability vector. Now suppose, on the contrary, that there are two recurrent states E_i and E_j, such that E_j cannot be reached from E_i. Then E_i cannot be reached from E_j either. If we assign the probability $1/r_{kk}$ to each state E_k in $\mathcal{R}_i$ and zero to all other states, we obtain a fixed probability vector. On the other hand, if we assign the probability $1/r_{kk}$ to each state E_k in $\mathcal{R}_j$ and zero to the states that are not in $\mathcal{R}_j$, we obtain a second fixed probability vector. Thus in this case there are at least two fixed probability vectors. (In fact, there are infinitely many such vectors, as is shown in Exercise 73.) In short, first, every Markov chain has at least one fixed probability vector. Second, there will be only one such vector exactly for those chains where we can get from each recurrent state to every recurrent state.

We summarize the main conclusions of this section. Suppose we have a Markov chain where every recurrent state can be reached from each recurrent state. Then there is a unique fixed probability vector. The entries in this vector are the fractions of the time that, in the long run, will be spent in each of the states.

Exercises

57. Find a fixed probability vector for

$$\begin{bmatrix} 1/2 & 1/2 & 0 \\ 3/4 & 0 & 1/4 \\ 0 & 1/4 & 3/4 \end{bmatrix}.$$

58. Find a fixed probability vector for

$$\begin{bmatrix} 0 & 1/2 & 1/2 \\ 1 & 0 & 0 \\ 0 & 1 & 0 \end{bmatrix}.$$

59. Find a fixed probability vector for

$$\begin{bmatrix} 1/2 & 0 & 1/2 \\ 0 & 1/2 & 1/2 \\ 0 & 1/3 & 2/3 \end{bmatrix}.$$

60. Find a fixed probability vector for

$$\begin{bmatrix} 0 & 1 & 0 & 0 \\ 0 & 0 & 1/3 & 2/3 \\ 1 & 0 & 0 & 0 \\ 0 & 0 & 0 & 0 \end{bmatrix}.$$

61. Find a fixed probability vector for

$$\begin{bmatrix} 0 & 0 & 1 & 0 \\ 0 & 0 & 0 & 1 \\ 1/2 & 1/2 & 0 & 0 \\ 1/3 & 2/3 & 0 & 0 \end{bmatrix}.$$

62. Find a fixed probability vector for

$$\begin{bmatrix} 0 & 1/3 & 0 & 2/3 \\ 1/3 & 0 & 2/3 & 0 \\ 0 & 1/2 & 0 & 1/2 \\ 1/2 & 0 & 1/2 & 0 \end{bmatrix}.$$

63. Find a fixed probability vector for

$$\begin{bmatrix} 0 & 1/2 & 0 & 1/2 \\ 1/2 & 0 & 1/2 & 0 \\ 0 & 1/2 & 0 & 1/2 \\ 1/3 & 0 & 2/3 & 0 \end{bmatrix}.$$

64. Find a fixed probability vector for

$$\begin{bmatrix} 1/2 & 1/2 & 0 \\ 1/2 & 0 & 1/2 \\ 0 & 0 & 1 \end{bmatrix}.$$

65. Consider the Markov chain with transition matrix

$$\begin{bmatrix} 1/2 & 0 & 1/2 & 0 & 0 \\ 1/3 & 1/3 & 1/3 & 0 & 0 \\ 0 & 1/2 & 1/2 & 0 & 0 \\ 0 & 0 & 0 & 1/2 & 1/2 \\ 0 & 0 & 0 & 1/3 & 2/3 \end{bmatrix}.$$

 a. Assuming a start in E_5, find the fraction of the time spent in each of the states.

 b. Assuming a start in E_2, find the fraction of the time spent in each of the states.

66. Consider the Markov chain with transition matrix

$$\begin{bmatrix} 1/2 & 0 & 1/2 & 0 & 0 & 0 & 0 \\ 0 & 1/3 & 0 & 1/3 & 0 & 1/3 & 0 \\ 1/2 & 0 & 1/2 & 0 & 0 & 0 & 0 \\ 0 & 0 & 0 & 1 & 0 & 0 & 0 \\ 1/2 & 0 & 0 & 0 & 1/4 & 0 & 1/4 \\ 0 & 0 & 0 & 1/2 & 0 & 1/2 & 0 \\ 0 & 0 & 1/2 & 0 & 0 & 0 & 1/2 \end{bmatrix}.$$

 a. Assuming a start in E_5, find the fraction of the time spent in each of the states.

 b. Assuming a start in E_2, find the fraction of the time spent in each of the states.

67. In the situation of Exercise 1, in the long run, what fraction of the time does each artist have a painting on display?

68. Do the last problem for the situation of Exercise 2.

69. In the situation of Exercise 3, in the long run, what fraction of the sales does each of the brands have?

70. In the situation of Exercise 5, in the long run, what fraction of the time does the sales representative spend in each of the cities?

**71. a. How could the answer to the last problem be found with a minimum of computation?

b. State and prove a proposition about this general situation.

72. In the situation of Exercise 6, in the long run, what fraction of the time does the man play each machine?

73. Suppose both

$$(a_1, \ldots, a_s) \quad \text{and} \quad (b_1, \ldots, b_s)$$

are fixed probability vectors for a certain Markov chain. Show that, for every number α with $0 \le \alpha \le 1$,

$$(\alpha a_1 + (1 - \alpha)b_1, \ldots, \alpha a_s + (1 - \alpha)b_s)$$

is also a fixed probability vector.

Table of Important Distributions

Name	k for which $P(X=k) \neq 0$	$P(X=k)$ for these k	$E(X)$	$Var(X)$	Generating function	Description
Bernoulli	$0,1$	q, p	p	pq	$pz+q$	One trial; 1 if success, 0 if failure
Binomial	$0,1,\ldots n$	$\binom{n}{k}p^k q^{n-k}$	np	npq	$(pz+q)^n$	Number of success in n Bernoulli trials
Geometric	$1,2,3,\ldots$	$q^{k-1}p$	$\dfrac{1}{p}$	$\dfrac{q}{p^2}$	$\dfrac{pz}{1-qz}$	Number of Bernoulli trials needed to get one success
Pascal	$r, r+1, \ldots$	$\binom{k-1}{r-1}p^r q^{k-r}$	$\dfrac{r}{p}$	$\dfrac{rq}{p^2}$	$\left(\dfrac{pz}{1-qz}\right)^r$	Number of Bernoulli trials needed to get r successes
Poisson	$0,1,2,\ldots$	$\dfrac{m^k}{k!}e^{-m}$	m	m	e^{mz-m}	Approximates binomial for n large, but $m=np$ not large
Hypergeometric	$\max(0, n-T+R) \le k \le \min(n,R)$	$\dfrac{\binom{R}{k}\binom{T-R}{n-k}}{\binom{T}{n}}$	$\dfrac{nR}{T}$	$\dfrac{n(T-n)}{T-1}\dfrac{R}{T}\dfrac{T-R}{T}$	complicated	Number of red balls chosen when n balls are chosen from T balls, R of which are red
Uniform	$1,2,\ldots,n$	$\dfrac{1}{n}$	$\dfrac{n+1}{2}$	$\dfrac{n^2-1}{12}$	$\dfrac{z^n+z^{n-1}+\cdots+z}{n}$	Choose one of the numbers $1,2,\ldots n$

Answers

In some cases, the answers below are approximations accurate to the number of places shown. In those cases, the number of places shown is somewhat arbitrary.

Chapter One

3a≫ 13/32   3b≫ 1/16   3c≫ 1/4   4a≫ 1/36   4b≫ 1/6
4c≫ 1/2   4d≫ 1/2   4e≫ 1/2   4f≫ 7/36   4g≫ 0
4h≫ 1/2   4i≫ 1/36   4j≫ 0   4k≫ 17/36   4l≫ 1/4
4m≫ 7/12   4n≫ 1/12   4o≫ 1/4   4p≫ 1/4   4q≫ 1/4
4r≫ 0

Chapter Two

1a≫ 20   1b≫ 6   1c≫ 24   2a≫ 40,320   2b≫ 2520
2c≫ 4,989,600   3a≫ 3,360   3b≫ 34,650   3c≫ $5.2797 \cdot 10^{14}$

3d≫ 3,087,564,480   4≫ 27,720   5≫ 3,360   6≫ 360

7a≫ 60   7b≫ 24   8≫ 88,080   9a≫ 8,640   9b≫ 518,400

10a≫ 24   10b≫ 12   11≫ 2,880   12≫ 4,838,400   13≫ 1,440

14≫ .13333   15a≫ $1.0218 \cdot 10^{20}$ 15b≫ $9.3667 \cdot 10^{18}$

16a≫ 40,320   16b≫ 5,040   16c≫ 384   16d≫ 96   16e≫ 96

17a≫ 720   17b≫ 576   17c≫ 8,640   19≫ 29   20a≫ 4,320

20b≫ 5,280   21a≫ 792   21b≫ 252   21c≫ 120   21d≫ 420

21e≫ 672   21f≫ 540   22a≫ 60,480   22b≫ 480

22c≫ 28,800   22d≫ 2,880   23a≫ 531,441   23b≫ 6,400

23c≫ 160,000   23d≫ 16,000

24≫ With $d = 54145$, we have $35673/d$, $16215/d$, $2162/d$, $94/d$, $1/d$

25≫ 1/3   26≫ .22857   27≫ .494949   28≫ .00264

29a≫ 3,003   29b≫ 60,060   30a≫ 756,756   30b≫ 126,126

30c≫ 6,054,048,000   31≫ 1/2

32≫ With $d = 4165$, we have   a≫ $1/d$ b≫ $6/d$ c≫ $88/d$

d≫ $1760/d$ e≫ $198/d$

33≫ 46,376   34≫ 1,330   35≫ 426,888   36≫ 111

37≫ 207,720   38a≫ 286   38b≫ 816   39≫ .34985

40≫ .01157   41≫ .09722   42≫ 495   43≫ $7.97 \cdot 10^{14}$

44≫ $2.41 \cdot 10^{13}$ 45≫ 3,386,880   46a≫ .212121

46b≫ .0242424   47≫ $6.9156 \cdot 10^{11}$ 48a≫ 15   48b≫ 6

48c≫ 30   49≫ 12,870   50≫ 2,159   51≫ 12,289

52a≫ $2.82 \cdot 10^{8}$ 52b≫ 8,008   53a≫ .001   53b≫ .006

53c≫ .003   54a≫ .0001   54b≫ .0024, .0012, .0006, .0004

55≫ $3.8719 \cdot 10^{-8}$, $1.1151 \cdot 10^{-5}$, $6.5512 \cdot 10^{-4}$, $8.3713 \cdot 10^{-4}$

56≫ $5.556 \cdot 10^{-2}$, $2.497 \cdot 10^{-3}$, $8.512 \cdot 10^{-5}$, $1.957 \cdot 10^{-6}$, $2.275 \cdot 10^{-8}$

Chapter Three

5≫ Yes, Yes, No   6≫ TFFF TTTF   7a≫ .125, .25, .3125, .3125

7b≫ .1552, .2688, .29952, .27648   8a≫ .015625   8b≫ .09375

8c≫ .1875   8d≫ .140625   9≫ 1/6   10a≫ .056314

10b≫ .001953   11a≫ .062500   11b≫ .197531   11c≫ .135031

12a≫ .401878   12b≫ .131687   12c≫ .270190   13≫ .038580

14≫ .169753   15≫ .15625   16≫ 1/2   17a≫ .38580

17b≫ .51775   17c≫ .09645   18a≫ .92981   18b≫ .92981
19≫ 1/2, 1/2, 1/2   20a≫ .6, .6   20b≫ *i/j*
21≫ .66510, .61867, .59735   23≫ 50, 50&51, 51, 51&52
24≫ 2, 2 ,2, 2&3, 3, 3   25a≫ 3   25b≫ 1   25c≫ 5   26≫ 0, 0,
0, 1   27≫ 5, 4   34a≫ 1/3   34b≫ 1/2   35a≫ .32810
35b≫ .51775   36a≫ .88571   36b≫ .42857   36c≫ .02857
36d≫ .03226   36e≫ .06667   37≫ .010526   38≫ .04082
39≫ .505263   40≫ .561404   41≫ 2/3   42≫ .13672
43a≫ 1/5   43b≫ 1/2   43c≫ 1/70   43d≫ 1/16   43e≫ 1/7
44≫ .46667   45≫ .28889   46a≫ .42857   46b≫ .12245
46c≫ .14286   47≫ .22857   48≫ .27778   49a≫ .05263
49b≫ .05263   49c≫ .89783   50≫ .75758   51≫ 1/3
52a≫ .59737   52b≫ .10614   53≫ .27083   54a≫ .6
54b≫ .3   54c≫ .5   54d≫ .6   55a≫ .34286   55b≫ .77143
55c≫ .44444   55d≫ .4   56≫ 1/3   57≫ .81818   58≫ .97561
59a≫ .31034   59b≫ .02200   59c≫ .00112   60a≫ .45455
60b≫ .29412   61≫ .9   62≫ .432432, .324324, .243243
63≫ .16976, .50419, .32604   64≫ .28205, .25641, .23077, .23077
65≫ 2/3   66≫ .51923   67≫ .6   68a≫ 3/4   68b≫ 3/4
68c≫ .84375   68d≫ 3/4, .95107   69a≫ .58333   69b≫ .80769

Chapter Four

1a≫ $.30   1b≫ $.30   2≫ $4   3≫ $.64, −.36   4≫ $20
5≫ $10   6a≫ 0, .5   6b≫ 1, 5.5   6c≫ 0, 12.5   6d≫ 0, .5
6e≫ 0, .5   6f≫ 1, 0   6g≫ 1, 1   6h≫ 1, .0099   6i≫ 1, 99
6j≫ 1, 999,999   6k≫ 1, 1   6l≫ 1, 1.6   6m≫ 1, 1.43
6n≫ 1, 1.75   6o≫ 1, 2   7a≫ 5/2   7b≫ 1/4   10a≫ −2.375,
2.7344   10b≫ 2.375, 2.7344   11a≫ 3   11b≫ 2   12a≫ 200
12b≫ 5,000   13≫ 7/3, 14/9   14≫ $4.80, 9.60   15a≫ 7
15b≫ 7   15c≫ 57   15d≫ 57   16a≫ 204   16b≫ 204
16c≫ 158,404   16d≫ 158,404   16e≫ 1 020   16f≫ 0
19≫ *X* + *Y*:6&2, *XY*:9&19   20≫ 4   22≫ 2, 1   23≫ 210,175
24a≫ 3   24b≫ 2   24c≫ 600   24d≫ 400   24e≫ 3

24f≫ .01   25≫ 1,250, 625, 1,563,125   26≫ 13   27≫ 40
28a≫ 16/3, 4/3   28b≫ 6   29a≫ 2, 4/3   29b≫ 30
30≫ 0, 5,000   31a≫ 7/2   31b≫ 4   31c≫ 35/12   31d≫ 2
31e≫ 15/2   31f≫ 59/12   31g≫ 14   31h≫ 77
32≫ 130, 5,650 if the unit is cents   33≫ 270, 975 if the unit is
cents   34a≫ 40   34b≫ 2,600   34c≫ 20   34d≫ 125,375
34e≫ 2,640   34f≫ 127,995   35≫ 60   36≫ 40, 4.89898
37≫ 2   38≫ 8   39≫ 15/8, 1.1094
40≫ Arranged as in question:

4 18   2
2   5   1
6 39   3
8 90 26

41a≫ 7/2   41b≫ 91/6   41c≫ 7   41d≫ 49/4   41e≫ 637/6
41f≫ 329/6   41g≫ 230.028   42a≫ 2   42b≫ 5   42c≫ 4
42d≫ 4   42e≫ 20   42f≫ 18   42g≫ 25   43≫ 1
44≫ $6,000.50 & a stop sign   45a≫ 19/9   45b≫ 2.1912
46a≫ 5/4   46b≫ 3.11   46c≫ 5/13   46d≫ 4.44   46e≫ 13/4
46f≫ 3.95   46g≫ 1   46h≫ 9.05   47a≫ 10.6   47b≫ 21.2,
31.8, 42.4   48≫ 13/3   49≫ $1   50a≫ $30.30   50b≫ $30.60
51≫ with replacement: 2.3312, without replacement: 53/21
52≫ 10.5   53≫ 12   54≫ 37.62626   55a≫ 6   55b≫ 3.5
56≫ 14/3   57a≫ 5/3   57b≫ 23/3

Chapter Five

9≫ 3.5   10≫ 3.5   11≫ 12   12≫ 3.45   13≫ 3.3758
14≫ 19/8   15≫ 59/30   16≫ 6.9   17≫ 283/60   18≫ 14
19a≫ $.26337   19b≫ $.70233   20≫ 16/11   21≫ $479.60
22a≫ 49/24   22b≫ 26/15   23≫ 1   24≫ 21,998,000.25
25a≫ .86397   25b≫ .40462   25c≫ .32718   25d≫ .39320
25e≫ 1.86397   25f≫ .04874   25g≫ .70588   25h≫ 1.10385
26≫ 104/9   27≫ 10   28≫ 710/17
29≫ with replacement: .27430656 (exactly), without replacement:
571/2205   30≫ 11.25   31≫ 3.15689   32≫ 35/12   33≫ 14/9
34a≫ 23/3   34b≫ 104/63   35≫ $3e - e^2$ (approx.)

Chapter Six

1$\gg$ .075816   2a$\gg$ .223   2b$\gg$ .442   3a$\gg$ .0498   3b$\gg$ .577
4$\gg$ .938   5a$\gg$ .135   5b$\gg$ .323   6$\gg$ 4.6   7$\gg$ .751, .215,
.0307, .00291, .000209   8$\gg$ .384   9$\gg$ .184   10a$\gg$ .5
10b$\gg$ .153   10c$\gg$ 182.5   10d$\gg$ 56   12$\gg$ 12   13$\gg$ 5
14$\gg$ 1   15$\gg$ 158   16a$\gg$ $4^n\sqrt{2\pi}$ 16b$\gg$ $\sqrt{2/(\pi n)}$
16c$\gg$ $2(16/27)^n/\sqrt{3}$ 17$\gg$ 56.05   18$\gg$ .0178   19$\gg$ $30 \cdot 10^6$
20$\gg$ .126   22$\gg$ .1359   23$\gg$ .8186   24$\gg$ .8186   25$\gg$ .0606
26a$\gg$ .6827   26b$\gg$ .9973   26c$\gg$ .0797   29$\gg$ .8664, .9987
30$\gg$ $7.6 \cdot 10^{-24}$ 31$\gg$ .0013

Chapter Seven

1a$\gg$ $3/(1 - 2z)$ 1b$\gg$ $1/(1 - 3z)$ 1c$\gg$ $1/(1 + 3z)$
1d$\gg$ $3/(1 - 3z)$ 1e$\gg$ $z/(1 - 3z)$ 1f$\gg$ $1/(4 - z)$
1g$\gg$ $3/(3 - z)$ 1h$\gg$ $2/(1 - z)$ 1i$\gg$ $2/(2 + z)$ 2$\gg$ $z/(2 - z)$
3$\gg$ $z/2 + z^2/4 + z^3/8 + z^4/8$ 4a$\gg$ $(z + z^2 + \cdots + z^6)/6$
4b$\gg$ $z/(6 - 5z)$ 5a$\gg$ 1, 1, 1, 1; $1/(1 - z)$
5b$\gg$ $1 - a_0, 1 - a_1, 1 - a_2, 1 - a_3; 1/(1 - z) - f(z)$ 5c$\gg$ $a_1, a_2, a_3,$
$a_4; [f(z) - a_0]/z$ 5d$\gg$ $0, a_0, a_1, a_2; zf(z)$ 5e$\gg$ $f(z)/(1 - z)$
5f$\gg$ $[1 - f(z)]/(1 - z)$ 5g$\gg$ $zf(z)/(1 - z)$ 5h$\gg$ $[1 - zf(z)]/(1 - z)$
6a$\gg$ $1/6$ 6b$\gg$ $1/4$ 6c$\gg$ $1/2$ 6d$\gg$ $5/12$
8a$\gg$ $(1/6)(z + z^2 + \cdots + z^6)$ 8b$\gg$ $1/4 + z/2 + z^2/4$
8c$\gg$ $z/24 + z^2/8 + (z^3 + \cdots + z^6)/6 + z^7/8 + z^8/24$
8d$\gg$ $(z^2 + z^4 + \cdots + z^{12})$ 9a$\gg$ $1/4$ 9b$\gg$ 2   9c$\gg$ 4
10a$\gg$ $3/2$ 10b$\gg$ $19/12$ 11a$\gg$ $1/2$ 11b$\gg$ $3/4$ 12a$\gg$ $1/6$
12b$\gg$ $e - 1$ 12c$\gg$ $4e - e^2 - 2$ 13a$\gg$ $10/3$ 13b$\gg$ $4/9$
14a$\gg$ $8/3$ 14b$\gg$ $16/9$ 15a$\gg$ $1/2$ 15b$\gg$ $1/2$ 16a$\gg$ .3240
16b$\gg$ .7616   16c$\gg$ 1.1816   17a$\gg$ L 17b$\gg$ $2L - L^2$
17c$\gg$ $1/\log 2$ 18a$\gg$ $0, 1/2, 1/4, 1/8, \ldots$ 18b$\gg$ $z^2/(2 - z)$
18c$\gg$ 3   18d$\gg$ 2   19a$\gg$ \$.97, 2.58, 3.55   22a$\gg$ 4
22b$\gg$ $21/4$ 22c$\gg$ 1   22d$\gg$ $5/9$ 22e$\gg$ $2 \log 2$
22f$\gg$ $9 \log 1.5 - 3$ 22g$\gg$ 4

Chapter Eight

2≫ 5/12   3≫ 5/8   4≫ 2/7   5≫ 3/13, 5/13, 5/13
6≫ 8/15, 4/15, 2/15, 1/15   7a≫ 2/3   7b≫ 8/9, 7/9
7c≫ 16/27   9≫ .0000169   10≫ .9802   11≫ .449
12≫ .8784   13≫ .01220   14≫ .1241   15≫ 1/3   16≫ .2471
17≫ 1/11   18≫ .9999977   19≫ .9610   20≫ .2968
21≫ .6517   22≫ .9997   26≫ 6   27≫ 4   28≫ 2.5
29≫ 43/13   30a≫ 8/3, 10/3   30b≫ 16/3   31≫ 30
33≫ 12.1   34≫ 7.8   35≫ 17.3   36≫ 50   37≫ 12.1
38≫ 1000   39≫ 160.0   40≫ 267.5   41≫ 549.8
42≫ 205,665   43≫ 11 days 14 hrs   44a≫ .0183, .4492, .6662,
.8686, .8898   44b≫ 48,993, 5,058.5, 1,336.0, 44.5, 10.6   45≫ 6
46≫ 14   47≫ 42

Chapter Nine

$$1\gg \begin{bmatrix} 2/5 & 2/5 & 1/5 \\ 3/5 & 1/5 & 1/5 \\ 3/5 & 2/5 & 0 \end{bmatrix} \quad 2\gg \begin{bmatrix} 0 & 2/3 & 1/3 \\ 3/4 & 0 & 1/4 \\ 3/5 & 2/5 & 0 \end{bmatrix}$$

$$3\gg \begin{bmatrix} .9 & .1 & 0 & 0 \\ .04 & .9 & .05 & 0 \\ 0 & .1 & .8 & .1 \\ 0 & 0 & .1 & .9 \end{bmatrix}$$

$$4\gg \begin{bmatrix} 0 & 1/2 & 1/2 & 0 & 0 & 0 & 0 \\ 1/4 & 0 & 1/4 & 1/4 & 1/4 & 0 & 0 \\ 1/4 & 1/4 & 0 & 1/4 & 0 & 1/4 & 0 \\ 0 & 1/2 & 1/2 & 0 & 0 & 0 & 0 \\ 0 & 1/2 & 0 & 0 & 0 & 1/2 & 0 \\ 0 & 0 & 0 & 0 & 0 & 0 & 1 \\ 0 & 0 & 0 & 0 & 0 & 0 & 1 \end{bmatrix}$$

$$5\gg \begin{bmatrix} 0 & 2/3 & 1/4 & 0 & 0 & 1/12 \\ 4/5 & 0 & 1/5 & 0 & 0 & 0 \\ 1/4 & 1/6 & 0 & 1/4 & 1/3 & 0 \\ 0 & 0 & 1/3 & 0 & 2/3 & 0 \\ 0 & 0 & 1/3 & 1/2 & 0 & 1/6 \\ 1/3 & 0 & 0 & 0 & 2/3 & 0 \end{bmatrix}$$

$$6\gg \begin{bmatrix} .1 & .9 & 0 \\ .45 & .1 & .45 \\ 0 & .95 & .05 \end{bmatrix}$$

$$7\gg \begin{bmatrix} .8 & .2 & 0 & 0 & 0 & 0 & 0 \\ .5 & 0 & .5 & 0 & 0 & 0 & 0 \\ .5 & 0 & 0 & .5 & 0 & 0 & 0 \\ .5 & 0 & 0 & 0 & .5 & 0 & 0 \\ .2 & 0 & 0 & 0 & 0 & .8 & 0 \\ .1 & 0 & 0 & 0 & 0 & 0 & .9 \\ 0 & 0 & 0 & 0 & 0 & 0 & 1 \end{bmatrix}$$

$8a\gg$ $\mathcal{R}_1 = \mathcal{R}_4 = \mathcal{R}_5 = \{E_1, E_4, E_5\}$; $\mathcal{R}_2 = \{E_1, E_2, E_4, E_5\}$; $\mathcal{R}_3 = \{E_3\}$;
$\mathcal{R}_6 = \{E_6\}$; $\mathcal{R}_7 = \{E_1, E_2, E_3, E_4, E_5\}$; $R : E_1, E_3, E_4, E_5, E_6$; $T : E_2, E_7$;
$A : E_3, E_6$ $8b\gg$ $\mathcal{R}_1 = \mathcal{R}_5 = \mathcal{R}_7 = \{E_1, E_5, E_7\}$; $\mathcal{R}_2 = \mathcal{R}_6 = \{E_2, E_6\}$;
$\mathcal{R}_3 = \mathcal{R}_8 = $ all; $\mathcal{R}_4 = \{E_4\}$; $R : E_1, E_2, E_4, E_5, E_6, E_7$; $T : E_3, E_8$; $A : E_4$
$8c\gg$ $\mathcal{R}_1 = \{E_2, E_3, E_4, E_5, E_7\}$, $\mathcal{R}_2 - \mathcal{R}_4 - \{E_2, E_4\}$;
$\mathcal{R}_3 = \mathcal{R}_5 = \mathcal{R}_7 = \{E_3, E_5, E_7\}$; $\mathcal{R}_6 = \mathcal{R}_9 = \{E_6, E_9\}$;
$\mathcal{R}_8 = \{E_2, E_4, E_6, E_9\}$; $R : E_2, E_3, E_4, E_5, E_6, E_7, E_9$; $T : E_1, E_8$; A: none
$8d\gg$ $\mathcal{R}_1 = \mathcal{R}_4 = \mathcal{R}_5 = \mathcal{R}_6 = \{E_1, E_4, E_5, E_6, E_8\}$; $\mathcal{R}_2 = \mathcal{R}_3 = \{E_2, E_3\}$;
$\mathcal{R}_7 = \{E_1, E_4, E_5, E_6, E_7, E_8\}$; $\mathcal{R}_8 = \{E_8\}$ $R : E_2, E_3, E_8$;
$T : E_1, E_4, E_5, E_6, E_7$; $A : E_8$ $9a\gg$ $4/5$ $9b\gg$ $3/8$

$$9c\gg \begin{bmatrix} 1 & 0 & 0 & 0 \\ 4/5 & 3/8 & 1/3 & 1/5 \\ 2/5 & 1/2 & 1/6 & 3/5 \\ 0 & 0 & 0 & 1 \end{bmatrix}$$

$10\gg$ $v_{12} = 0$, $v_{22} = 3/5$, $v_{32} = 4/5$, $v_{42} = 0$, $v_{13} = 0$, $v_{23} = 2/5$,
$v_{33} = 1/5$, $v_{43} = 0$ $11a\gg$ $3/5$ $11b\gg$ $7/12$

$$11c\gg \begin{bmatrix} 1 & 0 & 0 & 0 & 0 \\ 3/5 & 7/12 & 2/3 & 1/2 & 2/5 \\ 2/5 & 2/3 & 7/12 & 3/4 & 3/5 \\ 1/5 & 1/3 & 1/2 & 7/12 & 4/5 \\ 0 & 0 & 0 & 0 & 1 \end{bmatrix}$$

$12\gg$ $v_{12} = v_{52} = v_{13} = v_{53} = v_{14} = v_{54} = 0$, $v_{22} = 7/5$, $v_{32} = 8/5$,
$v_{42} = 4/5$, $v_{23} = 8/5$, $v_{33} = 7/5$, $v_{43} = 6/5$, $v_{24} = 6/5$, $v_{34} = 9/5$,
$v_{44} = 7/5$ $13a\gg$ $3/8$ $13b\gg$ $1/3$

$$13c \gg \begin{bmatrix} 1 & 0 & 0 & 0 \\ 3/8 & 1/3 & 1/3 & 5/8 \\ 1/8 & 1/3 & 1/3 & 7/8 \\ 0 & 0 & 0 & 1 \end{bmatrix}$$

$14 \gg v_{12} = v_{42} = v_{13} = v_{43} = 0, v_{22} = v_{32} = v_{23} = v_{33} = 1/2$

$15a \gg 1$ $15b \gg 1/2$

$$15c \gg \begin{bmatrix} 1 & 0 & 0 & 0 & 0 \\ 1 & 1/2 & 0 & 0 & 0 \\ 1/3 & 1/3 & 1/4 & 1/3 & 1/3 \\ 0 & 0 & 0 & 1 & 0 \\ 0 & 0 & 0 & 0 & 1 \end{bmatrix}$$

$16 \gg v_{12} = v_{42} = v_{52} = v_{13} = v_{23} = v_{43} = v_{53} = 0, v_{22} = 1, v_{32} = 2/3,$
$v_{33} = 1/3$ $17 \gg v_{15} = v_{25} = v_{35} = v_{45} = 0, v_{55} = 1/2$

$18 \gg v_{13} = v_{23} = v_{43} = v_{14} = v_{24} = 0, v_{33} = 1/3, v_{34} = 2/3, v_{44} = 1$

$19 \gg v_{11} = 3, v_{21} = 2, v_{31} = 0, v_{12} = 2, v_{22} = 1, v_{32} = 0$

$20 \gg v_{12} = v_{32} = v_{52} = v_{14} = v_{34} = v_{54} = 0, v_{22} = 1/8, v_{42} = 3/8,$
$v_{24} = 3/8, v_{44} = 1/8$ $21 \gg v_{11} = 1/2, v_{21} = 0, v_{31} = 0, v_{13} = 1,$
$v_{23} = 0, v_{33} = 1$ $22 \gg r_{12} = 2, r_{13} = 6, r_{23} = 4, r_{33} = 1$

$$23 \gg \begin{bmatrix} 5/2 & 3/2 & 3 \\ 1 & 5/2 & 4 \\ 2 & 1 & 5 \end{bmatrix}$$ $24 \gg r_{12} = 5/2, r_{22} = 1, r_{32} = 2$

$25 \gg r_{11} = r_{21} = r_{12} = r_{22} = 2, r_{31} = 8/3, r_{32} = 10/3, r_{41} = 4, r_{42} = 2$

$26 \gg r_{12} = 5, r_{13} = 2, r_{22} = 5/2, r_{23} = 2, r_{32} = 3, r_{33} = 5/3$

$27 \gg r_{11} = r_{22} = 1, r_{33} = 2, r_{43} = 2, r_{34} = 2, r_{44} = 2$

$28 \gg r_{11} = 5/2, r_{21} = 3, r_{12} = 2, r_{22} = 5/3$

$$29 \gg \begin{bmatrix} 7/3 & 2 & 12 \\ 8/3 & 7/2 & 10 \\ 20/3 & 4 & 7/2 \end{bmatrix}$$

$31a \gg$ 62 seconds, 34 minutes, 18 hours, 24 days, 2.1 years,
68 years, 697 centuries, 71 million years $31b \gg$ 2.6 hours,
2.3 years, 179 centuries, 139 million years $31c \gg$ 8.9 seconds,
31 seconds, 9.4 minutes, 15 hours, 60 days, 16 years, 15 centuries,
1,433 centuries, 14 million years, 1.3 billion years $32 \gg 73/10$
$33 \gg 216$ $34 \gg 222$ $35 \gg 216$ $36 \gg 216$ $37 \gg 43$ $38 \gg 4$
$39 \gg 10$ $40 \gg 8$ $41 \gg 8$ $42 \gg 7$ $43 \gg 7/2$ $44 \gg 9,331$
$45 \gg 46,656$ $46 \gg 83.2$ $47 \gg 80$ $48 \gg 5/2$ $49 \gg 5/3$
$50 \gg$ counting repetitions: 9.86, 9.43, 8.29, 9.86, 5.71, 0; counting

distinct rooms: 4.38, 4.32, 3.81, 4.38, 2.81, 0 (In each case the start is excluded.)   51≫ 77.5   52≫ 1, 2, 2, 1   53≫ 1, 2, 2, 1
54≫ 1, 5, 25/4, 9/2, 1   55≫ 1, 1, 3, 5   56≫ 1, 1, 1, 1, 10/3, 8/3
57≫ (3/7, 2/7, 2/7)   58≫ (2/5, 2/5, 1/5)   59≫ (0, 2/5, 3/5)
60≫ (1/3, 1/3, 1/9, 2/9)   61≫ (1/5, 3/10, 1/5, 3/10)
62≫ (3/14, 3/14, 2/7, 2/7)   63≫ (5/24, 1/4, 7/24, 1/4)
64≫ (0, 0, 1)   65a≫ 0, 0, 0, 2/5, 3/5   65b≫ 2/9, 1/3, 4/9, 0, 0
66a≫ 1/2, 0, 1/2, 0, 0, 0, 0   66b≫ 0, 0, 0, 1, 0, 0, 0
67≫ 1/2, 1/3, 1/6   68≫ 9/22, 4/11, 5/22
69≫ 1/5, 2/5, 1/5, 1/5   70≫ 6/29, 5/29, 6/29, 9/58, 6/29, 3/58
72≫ 19/75, 38/75, 18/75

Index

263

Undergraduate Texts in Mathematics

(continued from page ii)